Photographed Letters on *Wings*

How Microfilmed V-Mail Helped Win World War II

by Tom Weiner
with Bill Streeter

Copyright © 2017 Tom Weiner

All rights reserved, including the right of reproduction in whole or in part in any form.

Levellers Press, Amherst, Massachusetts
Printed in the United States of America

ISBN 978-1-945473-51-7

Dedicated to the memory of Henry Streeter,
who fought for his country in World War II
and gave his life before
he reached nineteen years of age.

Dedicated to the memory of Bill Streeter,
Henry's beloved cousin and V-Mail correspondent,
who died at age 86 in 2017 after living an extraordinarily full life.

Bill inspired this book and was dedicated to its completion.

TABLE OF CONTENTS

Preface	7
Introduction	12
Chapter 1 – *The Story of Microfilm*	24
Chapter 2 – *Airgraph – Its Origins, Uses, Successes and Challenges*	55
Chapter 3 – *V-Mail or Victory Mail: The Patriotic Way to Mail a Letter During World War II*	81
Chapter 4 – *Kodak and the Machines that Made V-Mail Work*	120
Chapter 5 – *V-Mail Stations Around the World*	136
Chapter 6 – *How Advertising Popularized V-Mail and How Artwork Made it More Appealing to Receive*	150
Chapter 7 – *The Voices of V-Mail*	161
Conclusion	188
Acknowledgments	195
Endnotes	198
Bibliography	207

Preface

By Bill Streeter

My name is Bill Streeter. I was born October 17, 1930 along with a twin sister, Wilma. It was the start of the Great Depression. My dad wrote in his diary the next day, "My favorite cow died last night. A set of twins was given us." In the Great Depression the loss of a farm animal was as dramatic as the birth of a child.

I will now introduce my double cousin, Henry Ward Streeter. I say double because his mother and my mother were sisters, and his father and my father were brothers. His mother died soon after his birth and Gramp and Grammy Wells brought him up. Their farm was next to our farm so he played with us most of the time. He really was more of an older brother than a cousin to me.

My generation was the kid brother of what Tom Brokaw called The Greatest Generation. We were just too young to go to war in the early 1940s. Henry was old enough to go to war in 1944, so he *was* one of The Greatest Generation.

The great American illustrator and painter, Norman Rockwell, documented our history on the covers of our magazines. So when *The Saturday Evening Post* came each week there we were. Today I call us "The Norman Rockwell Generation." We can recall standing with our pants down getting our first vaccination in the butt.

We were the kid sitting at the soda fountain with our bundle of clothing about to run away from home with the police officer nearby – two of the many Rockwell covers that depicted us.

Then there was the Thanksgiving dinner illustration, which was one of Rockwell's greatest, every detail of which tells the story of my generation.

There were a lot of lessons in the Great Depression. We didn't know we were poor. Mom and Dad never told us we were. I was the youngest of three boys in the family of five children and as the youngest you would know that when the package arrived from Sears and Roebuck with the back-to-school clothes in late August – there would be no new clothes for you.

You would not question that you were going back to school in hand-me-down knickers and shoes. You did not question anything.

You knew that was the way it was and you accepted life as it had to be.

We were a happy family. The farm was heavily mortgaged and we were a family of nine counting the hired men, but both Dad and Mom always had a chin up.

1932 was the height of the Depression with the stock market hitting a low of under fifty points and unemployment over twenty percent. But Dad wrote in his diary on the eve of 1933, "Good-bye old 1932. You were not so bad after all." His total cash flow that year was about $500.

President Franklin Delano Roosevelt opened up the 1940s with his Four Freedoms that should prevail everywhere in the world:

> FREEDOM OF SPEECH AND EXPRESSION
> FREEDOM OF WORSHIP
> FREEDOM FROM WANT
> FREEDOM FROM FEAR

These were the principles he wanted us to govern ourselves by.

Of course, our great American illustrator documented our four FREEDOMS in many different ways. So beautifully.

Then on December 7, 1941, we were told by Mr. Roosevelt that the Japanese had bombed Pearl Harbor. We were going to war.

We soon learned what war was all about. You learned patriotism and what deep love was. Yes, you also learned to hate. We learned our love of countr in the lyrics of songs like "The Yanks are coming and they won't be back 'til it's over over there." There was "My Filipino Baby" and "My Bonnie lies over the sea – bring back my Bonnie to me."

We learned to love in the many farewell parties we would have and we wondered would we ever see our soldier and sailor boy again.

We learned to hate from the walls of our town meeting place. They were covered with posters to remind us who our enemies were. I remember the one that showed three rats: one had Hitler's head to remind us of Germany and Nazism, one had the head of Mussolini to remind us of Italy and its fascism, and one had the head of Emperor Hirohito for Japan's imperialism. Today when I see a rat, I still see that poster and as a kid how we hated.

The Norman Rockwell Generation grew up in those war years. The Jewish culture has what they call a Bar Mitzvah when a boy is about to journey into manhood. We Yankee farm boys had our own

ritual for that journey. At age thirteen we were allowed to carry a jackknife. In the Sears Roebuck back-to-school package there would be a pair of fine leather Hi-Cuts and on the right leg the Hi-Cuts would have a sheath and in it was a jackknife.

That first day of school I was not the boy with hand-me-down clothing. I was the tallest kid on the playground with my leather Hi-Cuts and jack knife.

In late 1944 our own family member, Henry Ward Streeter, eighteen years old, was in a U.S. Army uniform. He had to go to basic training in the Carolinas for a few weeks. Then he came home for a furlough at Christmas before the long trip overseas in January 1945.

The U.S. Mail was the lifeline between family and soldier. Our family used V-Mail to write back and forth to Henry. It was faster –

twelve to fifteen days. We soon learned by V-Mail that he had been in the Battle of the Bulge and had received the Purple Heart. We learned by V-Mail that he was one of the last infantry foot soldiers over the Remagen Bridge before it collapsed into the great Rhine River.

Then history tells that in April his outfit, the 60th Infantry Regiment, 9th Infantry Division was shipped 160 miles due east to the Harz Mountains to liberate a concentration camp and a slave labor camp in one of the last great battles of World War II in Europe.

On April 18 the dreaded knock on the door came. Two U.S. military men had come to tell the family that Henry had been killed in action. Soon after the V-Mails started to be returned with "return to sender" written on the envelope. These words were followed by "Deceased."

It is more than seventy years later and you ask me, "Can you still love? Are you still patriotic?" The answer is yes. You ask, "Do you still hate?" In my imagination I see Henry lying in the cold spring mud in the mountains of Germany with the last drop of blood draining from a bullet hole in his chest. If only we could ask him if he still hates or if we could ask the corpses at the concentration camp if they still hate.

The one thing that we know, WE MUST NEVER FORGET.

Introduction

World War II has been presented through a variety of media in tremendous detail – from the battles and strategies to the lives of the soldiers and civilians who were so deeply affected around the globe. There have been books and films on a wide range of subjects over the seventy-plus years since the war ended. It would seem extremely unlikely at this point for there to be a subject related to the War that has not received sufficient treatment, and yet there is, at least one.

    I am referring to an essential aspect of the war that, since I learned about it, virtually no one I have spoken to is familiar: V-Mail. When I would tell friends and acquaintances about the book I've been working on, I'd hear, "E-mail? A book about e-mail? Everyone knows about it. Why waste your time on a book about such a well-known part of our culture?" "No, I said V-Mail." "What the heck is that?" That the word sounds so much like our widespread e-mail only adds to the misconceptions that surround the phenomenon. To complicate matters further, if you do a Google search leaving out the hyphen, which you can't include in speech, what pops up first is a website that features this definition, "A *vmail* account is a second email account just for your voicemail messages."

    V-Mail, (with its hyphen placed appropriately) originally called Victory-mail and at first written with the Morse Code for V as V..._ mail, was the means by which over a billion letters to and from the United States and at least half a billion to and from the United Kingdom, where it was known as Airgraph, were sent and received during the war years. Pioneered in England in the '30s, it was a system that employed microfilm and the machinery required to reproduce letters in miniature form so they could be shipped safely, securely and inexpensively to and from U.S. and British soldiers fighting throughout the world. When the microfilmed letters – 1,600 per roll – arrived at their destination they were enlarged and delivered to the recipient so they could be easily read.

In this day and age, where communication is instantaneous, it is hard to imagine the importance that written correspondence had for those fighting and those at home during World War II. Knowing that one's letters were going to reach their destination faster and more efficiently as a result of Airgraph and V-Mail was a morale booster of major proportions.

But it turns out that there is a deeper motive for bringing attention to this compelling though virtually unknown aspect of our history. Bill Streeter, man of many careers from farmer to psychiatric aide to businessman, from bookbinder extraordinaire to teacher of bookbinding, from local historian of his beloved Cummington, Massachusetts, to author, has a huge role to play in the effort to get the story of V-Mail told.

Bill began work on a book about V-Mail years ago and here's what he had to say when I interviewed him in preparation for writing this book, which he inspired:

> *The original motive to write this book comes from my cousin, Henry Ward Streeter, having been killed in action in Germany on April 17, 1945. The family had written Henry several V-Mails around April 17, and they were all returned weeks later. Stamped on the back of each letter in large red letters was the word "Deceased." Hence it has always been a desire to do a history of World War II V-Mail in memory of Henry.*

Of course, I never knew Henry, and Bill has written about their family connection and friendship in his eloquent and heartfelt preface. Bill invited me on a "roadtrip" to Cummington in late November 2016 where he'd arranged for me to have a chance to spend time with Henry in the only way available, which is described in a photograph of a letter I took that day: "Henry's step-mother, Florence Streeter, had begun keeping this scrapbook for him before his death and had

> Henry Ward Streeter was born in Cummington, Mass. May 13, 1926. His parents were Franklin W. Streeter and Ruth Wells Streeter. His mother died in 1929 and he was cared for by his maternal grandparents, Henry and Emogene Wells of Plainfield. He attended Plainfield schools and graduated from Sanderson Academy in June 1944.
>
> After induction into the Army in August 1944, Henry was sent to Camp Croft, S.C. for basic training where he qualified as Rifleman. His overseas assignment after January 1, 1945 was Co. A, 60th Infantry Regt., 9th Division, 1st Army. He was somewhere in Germany when he was killed by enemy action, April 17, 1945, only a few weeks before VE Day.
>
> Henry's step-mother, Florence Streeter, had begun keeping this scrapbook for him before his death and had saved the other material with which this book was completed by the Cummington Historical Commission in April 1987.

saved the other material with which this book was completed by the Cummington Historical Commission in April 1987" [forty-two years after his death].

It is important to remember that Henry, as was true of so many young men who fought and died in World War II, was a teenager when he was drafted. Here's the letter I discovered in the scrapbook

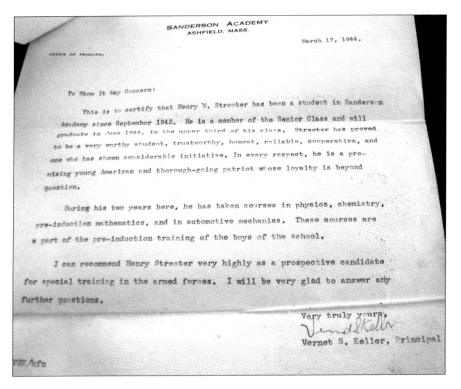

written by the principal of his school, Sanderson Academy in Ashfield, Massachusetts, attesting to his talents and his character. It mentions his being a "thorough going patriot whose loyalty is beyond question." It highlights his having taken courses as "part of the pre-induction training of the boys of the school," since by 1944 when it was written, the level of preparedness on the part of high schools of their male students was intense and systematized.

Upon graduating high school, Henry received his draft card and his pre-induction physical notification and had to report at the time and place it stipulated. Here are photographs of his high school graduating class. Henry is second from the right in the second row. His draft card and his notice to report for his physical are shown on the next page.

On page seventeen is a letter sent from Camp Croft, South Carolina informing family members of their son's arrival for four months of training in the Infantry, "a fine and important arm of the service." It emphasizes that the family can expect few letters given the rigors of the training program, but it also foreshadows what becomes standard

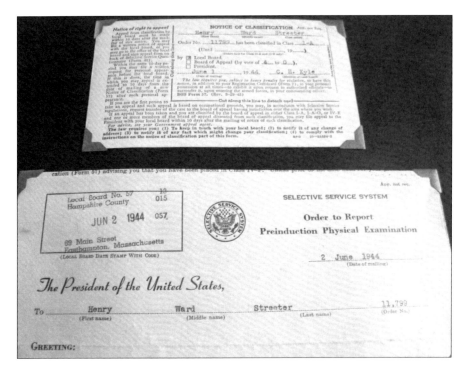

operating procedure; that their soldier "will be anxious for mail from home – and lots of it." As you will see in chapter seven, "The Voices of V-Mail," Henry's stepmother took this message to heart. It also reminds the families to "avoid delay and loss of mail by addressing your letters and packages correctly." It was 1944, after all, and such issues had been occurring for close to four years by the time Henry had been drafted. Henry received the training his peers were given and the scrapbook had a copy of the training manual (shown at left) with all of the varied skills that were taught during this no doubt harrowing time in his life.

Having a local boy get drafted, trained and furloughed necessitated an article in the local newspaper about the farewell party they threw for Henry prior to his departure for Fort Meade and soon after for Europe.

Bill described the farewell party for his double cousin in the preface and the photograph taken at that party is in the scrapbook and shown on page eighteen.

Having "successfully completed" a basic training course as a rifleman, Henry was awarded a certificate to honor his accomplishment.

As Bill tells us, Henry corresponded using the V-Mail system as did countless other soldiers. In the scrapbook I found several of the V-Mails he sent to family members in Cummington. Each is worthy of sharing because it tells about his experience on the Front and

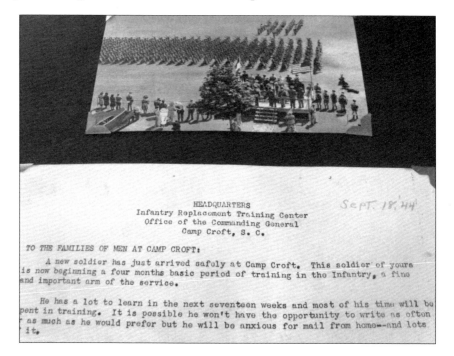

close to the end of his life. The first was written on January 26, 1945, and shows his desire to offer comfort to the recipient when he writes, "Don't worry." This is a principal concern of each of the letters. He is somehow able to give some details on the "outfit" and division he is in, perhaps a result of reduced censorship as the end of the war approached. Another theme that emerges is the universal one of asking for more letters.

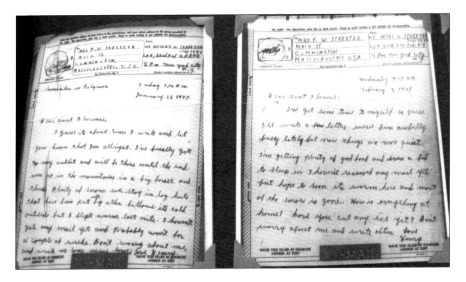

The next letter written on February 7, 1945, reiterates the two ideas that flow through each – "Don't worry about me and write often." He also asks his aunt whether they've cut any ice yet, clearly an act that he associates with home and the season and both of which were very much on his mind.

The first of the two letters from the scrapbook written by Henry during the month of March 1945 reveal that his unit was unable to receive mail for some period of time since he mentions his pleasure at receiving eighteen letters at once. He wrote, "I was glad to get them and hope there is a lot more coming."

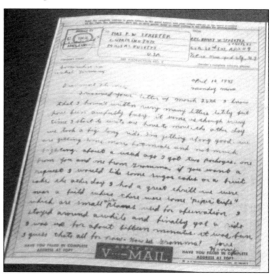

In another V-Mail letter Henry recognizes how little he's been able to write because, "We've been in the thickest of it and I did come out on top." He goes on to comment about the gap between his unit's current position and what his family member actually knows. "I know that you folks have been following me by the papers, but little do you realize exactly where I am, because the news you get are two or three days behind us." This is followed by the tragically prophetic words: "I am fine and well and hoping it will all end very soon." Seventeen days after writing this V-Mail, Henry was dead.

But he did write at least two more letters, the final ones his loved ones in Cummington received before his untimely death. In one he writes about moving too often to even write, but also about the thrill of being able to go up in a Piper Cub bi-plane, clearly a major treat for this soldier.

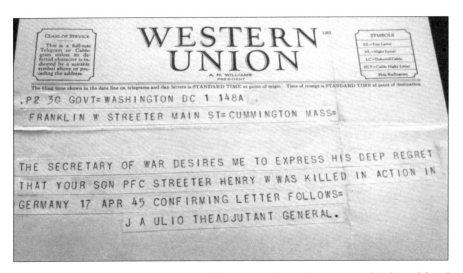

This was the telegram, which told of Henry's fate.

Senator Leverett Saltonstall wrote one of the "confirming letters" that arrived six weeks after the telegram:

To the left is is Henry's obituary.

Henry was posthumously awarded the Purple Heart. The official document alongside the Purple Heart is now on display at the Cummington Historical Museum.

A friend of Henry's family was able to go to the cemetery where Henry was buried in the Netherlands and share the photograph of his grave shone on the next page.

One final letter deserves to be included in this tribute to Henry's life and death. It was sent to Henry's step-mother by a woman who saw his obituary in the *Springfield Union* and wrote that she "felt so terribly sorry just as if I knew him personally." She goes on to tell about the tragic loss of her brother three months earlier. "My mother is a widow and along in years and we two are now alone to share the wonderful memories our grand soldier left behind…although we will grow old, they will remain the handsome, smiling heroes they were when they left home."

The V-Mail letter on the next page was written by Bradford (Brick) Liebenow of Cummington and dated April 16, 1945, the day before Henry died. It refers to Bill/Billy, his cousin who so inspired me with Henry's story and with the urgency of writing about V-Mail. The letter was returned in the envelope pictured along with the photograph of Henry with the following words on the envelope:

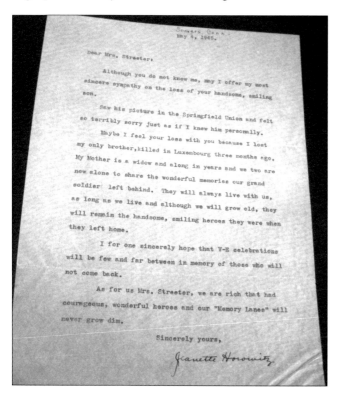

THE ENCLOSED V-Mail LETTER IS RETURNED BECAUSE IT WAS UNDELIVERABLE AT THE ADDRESS SHOWN OR TO WHICH FORWARDED

It said this on the picture:

Pvt. Henry Ward Streeter
Killed in Action
April 17, 1945

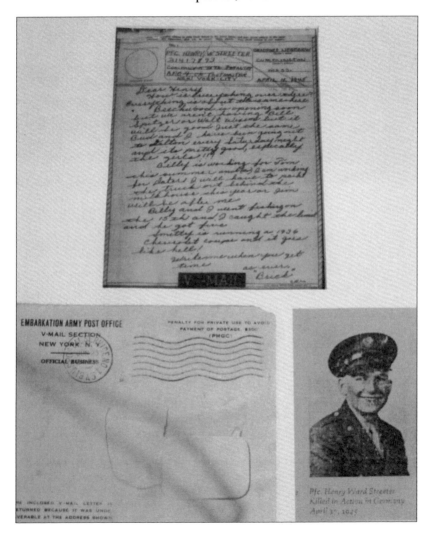

Not only is this book intended to honor Bill Streeter's fallen cousin, but to reveal Bill's careful research on the subject of V-Mail, with his discovery that there were numerous inaccuracies in the scant writing about it over the years. So here was purpose number two – to tell the true story of V-Mail and its role in enabling people to stay connected during the harrowing years of a tragic war that engulfed the world and had U.S. soldiers serving virtually everywhere on the planet. Rather than the inaccurate and distorted story that the whole project of V-Mail was unsuccessful, Streeter knew from personal experience that it had sustained relationships and provided invaluable information about loved ones. That V-Mail did so efficiently, reliably and cheaply adds to its stature as an enterprise that successfully combined the efforts of government and industry. That no one has until now written a book on the subject makes it even more essential that the true story be revealed.

As with my first book chronicling the testimonies of men and women confronted by the Vietnam War, this is a story that needs to be told and preserved. I am grateful to Bill Streeter for his pioneering efforts to get the story of V-Mail into the world. I am also indebted to Steve Strimer, publisher and true believer in the importance of giving voice to the voiceless, and Steve O'Halloran, beloved friend of Bill and champion of this book, for bringing the story to me.

CHAPTER ONE

The Story of Microfilm

If you are anything like me – meaning that you are at least fifty years old and have experienced microfilm through researching newspapers for high school report assignments – you'll no doubt join me in being mind-boggled to learn that microphotography began over 170 years ago.

If you are not at least fifty you might have little awareness let alone knowledge of microfilm, so I will begin this chapter's case study of microfilm's history with a definition that will get us all on the same page. The *Oxford English Dictionary* defines microfilm or microfiche as "film containing microphotographs of a newspaper, catalogue, or other document." The definition is enhanced by Ralph De Sola in his book *Microfilming*[1] in which he writes that microfilm is: "The reduction of images to such small size that they cannot be read without optical assistance." These two definitions combine to present the key ideas of reducing photographs along with the need to view them with some assistant technology.

Hence the researcher operating before the days of computers and online databases of newspapers, books and catalogues would often find him or herself in a library carrel, viewing microfilm using a microfilm reader. A check of the earliest newspapers available on microfilm reveals dates from the late eighteenth century. Yes, quite a resource to have at one's fingertips!

Now let's go back 178 years and chart the course of the version of film that eventually was responsible for the shipment of close to two billion V-Mail letters during World War II – letters that enabled soldiers and their families from Great Britain and the United States to stay connected in ways that bolstered the spirits of all.

The man known as the Father of Microphotography is John Benjamin Dancer, a British scientist. He began his experiments with and the manufacture of micro-produced novelty texts as early as 1839. Using a daguerreotype process, he succeeded in establishing a ratio of 160 to one. In 1853, Dancer successfully sold microphotographs as slides to be viewed with a microscope.[2]

Dancer's next achievement was the result of the pioneering work of Frederick Scott Archer, a fellow Englishman. Dissatisfied with the calotype of photography that resulted in poor definition and contrast and the need for long exposure time, Scott Archer invented the wet collodion process in 1848. The collodion process, mostly synonymous with the "collodion wet plate process," required the photographic material to be coated, sensitized, exposed and developed within the span of about fifteen minutes, necessitating a portable darkroom for use in the field. This enabled photographers of the time to combine the fine detail of the daguerreotype with the ability to print multiple copies like the calotype.[3]

Archer knowingly published his discovery in *The Chemist* in March 1851 without seeking a patent. He wanted it to be his "gift to the world."[4] He is remembered for this singular achievement, which Dancer perfected with his microphotography endeavors. Archer died in obscurity and poverty because of his decision not to patent his work. Dancer was one of the beneficiaries of his efforts. He continued his fascination and experiments with microphotography and the wet collodion process of Archer, but he himself diminished the importance of his own work. Rather than see the potential that microphotography offered, Dancer dismissed his own decades-long work as a personal hobby. As a result he did not document his procedures.

In 1851, James Glaisher, an English meteorologist and astronomer, suggested that storing documents could be a way to make use of microfilm. He is best known for his pioneering work as a balloonist.

He sought to measure the humidity and temperature at the highest possible reaches of the atmosphere. He broke the world record for altitude on September 5, 1862, but before a final determination of his altitude could be taken, he passed out at 8,800 meters. One of the several pigeons accompanying him on the ascent unfortunately died, but within the next eight years, pigeons were to play an essential role in what eventually became V-Mail.[5]

The photography exhibit at the Great Exhibition in London in 1851 had a major impact on Glaisher. Following the exhibition he called photography "the most remarkable discovery of modern times," arguing in an official report for using microphotography to preserve documents.[6] Soon after Glaisher offered his groundbreaking ideas about the potential uses of microphotography, Sir John Herschel took up the charge.

Herschel originated the use of the Julian day system in astronomy, named seven moons of Saturn and four moons of Uranus, made many contributions to the science of photography, and investigated color blindness and the chemical power of ultraviolet rays. His *Preliminary Discourse* (1831), which advocated an inductive approach to scientific experiment and theory building, was an important contribution to the philosophy of science.[7]

In 1853 Herschel recognized the possibility of creating microscopic editions of reference works, pocket-size notes and manuscripts. He imagined reducing larger, bulkier materials to more manageable size for use by librarians and researchers, foreshadowing microfilm versions of newspapers. He also envisioned the use of the microscopic lens to read works produced by microphotography.[8]

With such influential and enthusiastic practitioners and visionaries one might have expected microphotography to be destined for some major breakthroughs in the world of photographic imaging, but the 1858 *Dictionary of Photography* deemed that microphotography had little utilitarian value. The *Dictionary* even went so far as to call the process "somewhat trifling and childish."[9]

The *Dictionary* was by no means the final word on the subject as the *Photographic News* for both 1859 and 1860 heralded microphotography and offered numerous suggestions akin to those already mentioned by Glaisher and Herschel, but going beyond their work to explore military applications.[10]

In searching for purveyors of microphotography during this time period and in the wake of such dismissive descriptors as emanated from the *Dictionary*, Alfred Reeves featured prominently in an anonymously written article on microphotographs that was reprinted in numerous magazines in 1859. Regarding Reeves and his work:

> Mr. Alfred Reeves has recently forwarded to us a specimen of one of those minute pictures, which consists of a plate containing the portraits of kings and queens of England since the time of the Conquest. Here, on a space not larger than 1/16 of an inch square, may be perceived a miniature "National Portrait Gallery" with a portrait of every king and queen

surrounding her Majesty, who is properly made the center figure of the interesting group.[11]

Next up in the development of microfilm as a viable means of transmitting records and improving communication is the French photographer and inventor, René Dagron. Once again Mr. Dancer features prominently as it was his exhibit of microfilm in Paris in 1857 that captured Dagron's attention. He immediately appreciated its potential, but saw the problem caused by having to view the microphotographs through a microscope, too costly for most people at the time.

To solve this dilemma, Dagron built upon the Stanhope lens, a simple one-piece microscope that had been invented earlier in the nineteenth century by Charles, the third Earl of Stanhope. The Stanhope lens is a cylinder of glass with each end curved outwards with one lens more convex than the other. Its simplicity and affordability made it popular. It was found to be quite useful in the world of medicine, enabling the viewer to examine such transparent materials as crystals and fluids.[12]

It was upon this lens that Dagron made his modifications. He kept one curved end to refract light while causing the other end to be flat and locating it at the focal plane of the curved side. He used this new lens to mount his microphotographs in what were called *bijoux photo-microscopiques* or "microscopic photo-jewelry."[13]

Dragon sought and was finally granted the first microfilm patent in history in 1859, the same year he introduced his photographic miniature Stanhope toys and jewels during the International Exhibition in Paris.[14] In 1862 Dagron returned the favor Dancer had offered and received an honorable mention at the London Fair. He presented a set of microfilms to Queen Victoria. That same year Dagron published his book: *Cylindres photo-microscopiques montes et non-montes sur bijoux, brevetes en France et a l'etranger*. (Translated as: *Photo-microscopic cylinders mounted and non-mounted on jewels: Patents in France and abroad*).[15]

Dagron's success and ensuing fame allowed him to build the first factory dedicated to the production of devices used to view microphotographs. Starting in 1859 his factory began producing the Stanhopes, which were mounted in jewelry and souvenirs. By 1862 Dagron's factory had 150 employees and was manufacturing 12,000 units a day. Soon afterward he launched a mail order marketing campaign to promote the viewing devices.[16]

In 1864, two key events propelled Dagron's fame. He produced a Stanhope optical viewer, which enabled the viewing of a microphotograph of one square millimeter, equivalent in size to the head of a pin. The photograph included the portraits of 450 people. The same year he published the thirty-six page booklet *Traite de Photographie Microscopique* in which he described in minute detail the process he invented to produce microfilm positives from normal size negatives. Given the wide array of his accomplishments it is no surprise that the microfilm industry is considered to have been created by him.[17]

Meanwhile there were other remarkable ideas floating around Dagron's newly patented invention. He himself saw the potential for alternative uses and actually complained that most writers on the subject of microfilm seemed to lack imagination and were blind to its possibilities. Ironically he pointed out that one of the major reasons for the limited view of those writing about microphotography resulted from his use of the medium for novelty souvenirs.

Dagron was not only a brilliant inventor, but a great marketer as well. One of his well-known ploys to generate "buzz" on the street, was to surreptitiously plant one of his jewelry pieces, a most unusual ring, on the Champs-Elysées. We have no record of the finder, though given Dagron's propensity for thoroughly planning such escapades, it could very easily have been someone in his pay. Whoever it was brought the ring to reporters assigned to the area. Looking into a tiny peephole, the reporters discovered an image.

The next day, Paris newspapers featured multiple stories about the new invention that would enable men to "gaze at their beloveds privately." Of the invention, the reporters wrote, "Nothing could be more extraordinary…than to find in the setting of a ring…a portrait…the size of a carte de visite."[18]

The publicity most definitely boosted sales of Dagron's product. It soon became evident that the device would be a splendid means of advancing the local pornography industry. Dagron's factory produced the device *and* the content for viewing. To stifle competition when it inevitably surfaced, Dagron and Company filed lawsuits, one of which sought damages from the Martinache Company. The lawsuits' failure led to the merger of Martinache and Dagron, but Dagron kept up the court battles eventually losing a bitter patent battle with a group of opticians who sold a micro-viewing machine similar to the one he'd invented. Despite his defeat, Dagron sought damages that would have allowed him to put forth some embarrassing publicity, but the judge dismissed his plea and required Dagron and Company to pay all court costs.[19]

Examples of other uses for microphotography being considered at this same time included developing an application to assist spies in wartime. Microfilm would provide a means for them to carry large numbers of documents in minute form lowering the risk of discovery even after capture and a thorough search. The thought was there would be almost no limit to the amount of material that could be contained in a single button or in the ornamental top of a pencil. Another suggestion, this time from *Photographic News*, was to place sensitive microfilmed documents inside a hollowed out bullet and then shoot the information over enemy occupied territory.

Little did the originator of this last idea realize how close the next development would be to his concept. He only missed the delivery method. It would not be a bullet, but rather a pigeon!

The use of pigeons as a means of transmitting data have biblical antecedents going all the way back to Noah who sent one skyward from his Ark to seek dry land following the Flood. Caesar used pigeons to conquer Gaul. They were used by the Greeks to transmit names of winners of the original Olympic Games. There's evidence that the first pigeon post actually occurred in ancient Baghdad in 1,150 where merchants used pigeons to transport information to one another. In 1860, Paul Julius Reuters, founder of the eponymous wire service, used pigeons to convey news and stock prices between Brussels, Belgium and Aachen, Germany.[20]

Speaking of the stock market, the Rothschild family succeeded in making a fortune thanks to a pigeon delivery of exceedingly timely news. Were Napoleon to win at Waterloo in 1815 British debt would be devalued whereas a British victory would cause the value of British debt to rise. Huge stakes to be sure![21]

The Rothschild family had been using pigeons to communicate throughout their vast empire. Nathan Rothschild received news of the British victory in London, watched British debt be undervalued and BOUGHT! Even the British didn't find out about the victory until the next day.[22]

Microfilm came of age during the Siege of Paris that occurred during the Franco-Prussian War. Much has been written about the war, some suggesting that, "both the siege and the war have been the subjects of a vast literature which is said to exceed that of any other historical event."[23] Remarkably, just six weeks after the outbreak of hostilities, Emperor Napoleon III and the French Army of Chalons surrendered at Sedan on September 2, 1870, resulting in the fall of the Second Empire and the advance of the Prussian army on Paris. The army of Napoleon III "went to war ill-equipped, badly led, trained and organized, and with inferior numbers."[24]

Within days of the declaration of the Third Republic on September 4, 1870, the fate of Paris became clear – there would be a siege – which necessitated moving the center of the French government to Tours on September 12. Obviously there still needed to be communication between Paris and Tours once the siege began, so a telegraph cable was obtained from England and secretly laid along the bed of the Seine between Paris and Rouen.

The government in Tours decided to consolidate the positions of director of the telegraph and director of the postal service into one, which was headed by Francois-Frederic Steenackers, born in Belgium, who became a naturalized French citizen in 1869. He was a deputy from the Haute-Marne in the lower house of Parliament in the Second Empire and had been influential in the expulsion of the Bonapartists. In addition to laying the hidden cable that maintained communication, Steenacker took a number of carrier-pigeons with him to Tours. By September 20, Paris was encircled and the Prussians had cut the normal channels of correspondence.[25]

The situation continued to deteriorate. The last overhead telegraph wires were cut on the morning of September 19, and the secret telegraph cable in the bed of the Seine was located and cut on September 27. Postmen attempted to deliver the mail, but few succeeded. Others were captured and shot. By the end of October there is no evidence of any post reaching Paris from outside the city except for private letters carried by unofficial individuals.[26]

Attempts were made to get mail into and out of the capital. Five sheepdogs that had previously driven cattle into Paris were flown out of the city by balloon in the hope they could return carrying mail, but after they were released they disappeared. Another effort featured zinc balls filled with letters and floated down the Seine, but again the results were dismal. Later it was said, *"Pas qu'une souris pût franchir les lignes prussiennes sans être vue."* (Not a mouse could cross the Prussian lines without being seen.)[27]

Parisians were completely isolated by the end of September. Anticipating this it had been suggested on September 2, based on the time-honored tradition of the role of carrier-pigeons, that all pigeons in Paris should be sent away and made ready to bring messages into the city while pigeons would be brought in from the north of France to be ready to carry messages out of the city. During the course of the siege pigeons were regularly taken out of Paris by balloon, but when paper messages were attached to the birds, the weight was too much and many fell into enemy hands. Only a few ever reached Paris.[28]

Several individuals and an organization deserve credit for what came next. Though a number of people tried to influence the government to use the pigeons, it was an influential Parisian lawyer to whom credit is due, Mr. Segalas. He succeeded in persuading the outgoing Director of Telegraph to erect a pigeon loft in Paris, though it could only have served as a staging post for pigeons being taken out of Paris and not one to which they would return since there would have been no opportunity to train pigeons to operate from it.[29]

The donation of the Parisian pigeons gets credited to L'Esperance, the Parisian pigeon fancier's club. Several of their members who transported birds by balloon risked capture and imprisonment to do so. An effort to leave the besieged city by train failed when it took too long to get official French approval. The Prussians had cut the rail lines by the time approval arrived.[30]

The pigeons were taken to their base after their arrival from Paris and when they had preened themselves, been fed and rested, they were ready for the return journey. Tours lies about 200 km from Paris and Poitiers some 300 km, as the crow or pigeon. flies. To reduce the flight distance, the pigeons were taken by train as far towards Paris as was safe from Prussian intervention. Before release, they were loaded with their dispatches.[31]

The first dispatch was dated September 27 and reached Paris on October 1. It consisted of paper tightly rolled and tied with a thread and then attached to a tail feather of the pigeon. It was necessary to make sure it wasn't an old feather, which could have been shed during the flight. The dispatch was protected by being inserted into the quill of a goose or crow and it was the quill that was attached to the tail feather.[32]

All was finally in readiness for the final piece of the puzzle involving pigeons and microfilm. Voilá – the reappearance of M. Dagron! At the Exposition Universelle of 1867 in Paris, he had demonstrated a remarkable standard of microphotography, which he had described in *Traite de Photographie Microscopique*. He now proposed to Postmaster General Germaine Rampont-Lechin that his process should be applied to pigeon messages. Minister of Finance, Ernest Picard, declared Dagron the *"chef de service des correspondances postales photomicroscopiques."* M. Dagron had sought a more profound use for microphotography than novelties and postcards. He had found it!

Pigeons crossed enemy lines undetected.[33]

Arrangements were made for Dagron to leave Paris by balloon, accompanied by colleagues, Albert Fernique; professor of engineering, Jean Poisot; artist and son-in-law of Dagron; Gnocchi, Dagron's assistant, and Pagano; the pilot. They departed on November 12th in the balloons Niépce and Daguerre, but the latter, with much of the equipment and pigeons in it, was shot down by the Prussians using a long-range, breach-loading artillery rifle used four years previously in the Battle of Sadowa, the battle that ended the Seven Weeks' War of 1866. "The first casualties of industrial warfare fell from the sky with one push of a button."[34]

The balloon fell within the Prussian lines, but not all members of the Daguerre crew were lost. From the wreckage of the Daguerre, six pigeons would now be called to battle. These pigeons were released to convey microfilmed news to Paris. Six identical messages were sent on each bird, "Large blue and yellow balloon fell at Joissigny. Prussians captured balloon, voyagers. Have been able to save a mailbag and six pigeons."[35]

The Niépce was more fortunate, but some of the precious photographic equipment had to be jettisoned in order for it to gain altitude and escape the fate of the Daguerre. It reached Tours where Dagron was able, despite the handicap of lost equipment, to set up the first microphotography unit ever employed in a war.

Of course, there were additional obstacles beyond the ones faced by our fellow humans trying to escape the city by balloon. Weather proved to be a deterrent for the pigeons as it became increasingly cold. The siege continued into the winter, one of the most bitter in memory, but it was not the only hazard they faced. There were their natural enemies, hawks, and their human predators, Frenchmen who with their shotguns were seeking food for their families as the war wreaked havoc with the French economy and agriculture. A total of 302 largely untrained pigeons left Paris during the course of the siege, and 57 returned to the city. Many also fell prey to Prussian rifles or the falcons that they hastily introduced to intercept France's feathered messengers.[36]

Rumors blamed some of the disturbance in the pigeon's homing sense on the noise of cannon fire, but this is false. What did happen was that the finest pigeons with the best homing sense and therefore the most reliability were used first and over time the birds employed would have received less training and so were increasingly less likely to return safely with their cargo to Paris.

The pigeons carried two kinds of dispatches – official government documents and private correspondence. Initially the service sent only official dispatches, but in early November it was opened to the public. We will see a similar progression later on with England's Airgraph system in the 1930s. Private letters were only sent along with official communiqués since the latter received absolute priority. It was Dagron's microfilms that eased any concerns about there being sufficient room for some combination of the two forms of correspondence. Their volumetric measurements were so small that, for example, one tube sent aboard one pigeon during January 1871 contained twenty-one microfilms, of which six were official dispatches and fifteen were private letters.[37]

The earliest messages were hand-written, but it was soon seen to be more efficient to use typeset characters in letterpress with the work done by printers and then photographed first in Tours until December 10, 1870, but then in Bordeaux when the advancing German army threatened Tours. Work on official dispatches began at noon and ended at 5:00 p.m. after which the private messages became the focus. As with the earlier messages, the microfilms were placed in goose-quill tubes and later in light-weight metal tubes, one-and-a-half- to two-inches long and tied to the bird's main tail feathers by strong, silk thread. Each pigeon was able to carry between twelve and eighteen of these pellicules as they were called, with the total weight less than one gram. Three thousand messages could be sent at a time![38]

The system was so refined that in order to improve the chances of a successful flight reaching Paris, the same dispatch was sent by several pigeons – up to 35 times for an official communiqué and on average 22 times for a private dispatch. Records indicate that from January 7 to the end on February 1, 61 tubes were sent off, containing 246 official and 671 private dispatches.[39]

The method of choice that was used involved sending off the dispatches not only by pigeons of the same release, but also pigeons of subsequent releases until Paris signaled a safe arrival. When the pigeon reached its specified loft in Paris, a bell in the trap in the loft was rung to announce its arrival. A watchman immediately removed the tube and it was taken to the Central Telegraph Office where its contents were unpacked and placed in a basin of water, mixed with a little ammonia, which caused the film to unroll so it could be dried and placed between two thin sheets of glass, which formed a transparency.[40]

An illustration from *The Graphic* of February 4, 1871 : "A Welcome Visitor – Arrival of a Pigeon in Paris".

The next step involved projecting the miniature images onto a screen or white wall using a Megascope, a type of magic lantern device invented in 1851 by Jules Duboscq, which employed the intensely bright electric arc lamp, an improvement upon the original arc lamp invented by Humphrey Davy in the first decade of the 1800s. The resulting enlargement, magnified 160 times, could be easily read and copied by a team of clerks. The transcribed messages were written out on forms (for private messages telegraph forms were used, both with or without the special designation, "pigeon") and then delivered.[41]

Many factors affected the time between the sending and receiving of the pigeon post dispatch. Among them were: the density of telegraphic traffic to and from the sender's town, the time taken to register the message, to send it to the printers where it was assembled with 3,000 other messages into a single page and then to assemble the pages into nines, twelves or sixteens. To the great credit of the pigeon flyers, during the siege's four months, 150,000 official and 1 million private communiqués were carried into Paris – the equivalent of a library of five hundred books![42]

There is some controversy as to the speed of the Pigeon Post. The first private messages did arrive quickly at their destinations. One excellent example pertains to an order Dagron made for some chemicals from a firm in Paris. A pigeon released on January 18, near

An engraving showing the projection of the microfilm by the Duboscq Megascope.[42]

Poitiers, flew the 185 miles to Paris in twelve hours. The chemicals were assembled and left the city aboard the balloon "General Bourbaki" at 5:15 a.m. on January 20. The balloon arrived in occupied territory near Reims at 2:00 p.m. It was immediately burned to prevent it from being captured, but its pilot, Theodore Mangin, delivered the chemicals four days later – perhaps even faster than in peacetime. Nevertheless, the variables outlined above – time of day, weather conditions, number of dispatches and the pigeon's skill, experience, age and condition – could result in a message that, from the handing in to its delivery, could easily span two months.[43]

The Pigeon Post service officially ended on February 6, 1871, on instructions provided by the Director-General of Posts. What followed led to some serious accusations of profiteering by members of L'Esperance. The surviving pigeons were sold at public auction for as little as three francs each, a mere fraction of the 36,000 francs negotiated by the lawyers on behalf of the club members. Several unnamed members were believed to have sold inferior pigeons to the postal service at inflated prices. Those birds whose wings were imprinted with their owner's name and corresponding serial number were returned without added expense being incurred.[44]

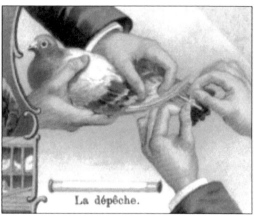

Affixing a pellicule to a pigeon's tail feathers

There was great appreciation for the role the balloonists and the pigeons played in making the siege more bearable. The Pigeon Post was definitely one of the very few successful aspects of the Franco-Prussian War as far as France was concerned. The arrival of a carrier pigeon with news from the world beyond Paris was a major morale booster for the city's residents who were suffering on so many levels from the Siege. It was even proposed that pigeons be added to the Paris coat of arms, but this was not to be.

Determined to honor their contribution, the pigeons' courage along with that of the aeronauts were commemorated with the issuance of medals and eventually in 1906 with the erection of a fine bronze monument. Funded by public subscription and designed by Frederic-Auguste Bartholdi, who earlier in his career had designed the Statue of Liberty, the statue was unveiled at the Porte des Ternes in Neuilly. Four pedestals, each one topped by a pair of bronze pigeons surrounded a central representation of a hot air balloon. For reasons we can imagine regarding the contribution of the Pigeon Post to safeguarding morale during their attempted siege, the Germans took offense at the monument and destroyed it in 1944 during the occupation of France.[45]

The success of the carrier pigeons during the Siege of 1870–71 established that these birds could make communication possible in the worst of circumstances. By 1899 Spain, Russia, Italy, France, Germany, Austria and Romania had each created their own pigeon services. Britain was thereby alarmed and "a call to arms published in the influential journal *The Nineteenth Century* expressed concern at the divergence in military capability" as represented by the pigeons. There was even a fear that, "The Empire... was being rapidly outpaced by foreign technology"!

The pigeons would once again prove their usefulness in later wars, but aircraft would diminish the centrality of their role by the time of World War II. It is still quite remarkable to think that one of the oldest forms of communication – the carrier pigeon – joined forces with one of the more modern forms of communication – microfilm – to enable the Siege of Paris in 1870 to be at least partially broken.

With the end of the Siege and a period of relative tranquility in Europe the next significant development in the microfilm story involves none other than George Eastman of Kodak Co. The Pigeon

Post microfilm used a gelatin emulsion that had to be stripped from the glass plates. This was most inconvenient, time consuming and inefficient. Eastman came up with a method that involved coating the paper with a layer of plain, soluble gelatin and then with a layer of insoluble light sensitive gelatin. After exposure and development, the gelatin with the image was "stripped" from the paper, transferred to a sheet of clear gelatin and varnished with collodion, a cellulose solution that forms a tough, flexible film.[46]

A postcard of the Monument to the Aeronauts and Pigeons of the Siege of Paris

With this process, which Mr. Eastman patented in 1884, he positioned his Kodak Co. as the leading contender in the development and creation of cellulose acetate film for the worlds of still photography, moving pictures and microfilm, but this was just the beginning of Kodak's microfilm related inventions of which more will be revealed soon.[47]

During the next few decades, microfilm did not achieve the recognition its role in the Franco-Prussian War appeared to indicate was in store. Instead there were a number of individuals who proposed usages that didn't get realized. In 1896, a Canadian engineer, Reginald A. Fessenden suggested that microfilm could be a solution to his fellow engineer's unwieldy, but regularly consulted materials. His proposal – put 150 million words in a square inch so that a one-foot cube could contain 1.5 million volumes. It remained a proposal.[48]

Ten years later, in 1906, two men, Paul Otlet, a Belgian author, entrepreneur, visionary, lawyer and peace activist and his countryman Robert Goldschmidt, chemist, physicist, and engineer, sought to make use of microphotography to make library materials more readily available. They both knew that "access to the libraries is not always easy and delays in the transmission of books often discourages the most tenacious workers, to the detriment of scientific progress.... Travel by scholars, the international exchange of scientific books between libraries, the copies or extracts requested from abroad, are seriously under-resourced." Goldschmidt subsequently wrote that there is a need for "a new form of book that will help overcome these major inconveniences." They determined that the solution lay in microphotography and proceeded to explain how "a single card measuring 12.5 x 7.5 cm, providing 72 square cm of space (margins excluded) could contain the contents of an entire 72-page book."[49]

In the 1920s, Otlet and Goldschmitdt proposed the creation of a World Center Library of Judicial, Social and Cultural documentation employing microfilm since it could offer a durable and inexpensive, easy to reproduce and extremely compact form. By 1925 the team was envisioning a massive micro-photographic library in which every volume existed as master negatives and positives. The library would consist of "pocket sized" viewing equipment, a portable cabinet about one meter wide and one meter high and ten centimeters deep, capable of handling 18,750 books of 350 pages each, equal to the books that would fill 468 meters of conventional library shelving. Upon learning of this proposed undertaking, I was reminded of the rumor that circulated in the wake of the 1974 opening of the twenty-six story UMass Amherst Library, still the tallest in the world, that all of its books could be copied onto CDs, which could contain all of the information of the entire library on one shelf! Once again, the idea remained unfulfilled, though the 1920s would see an explosion of similar ideas coming to fruition.[50]

With the outbreak of war in 1914, this time a world war referred to until World War II as the Great War, microphotography found another purpose, one that had actually been imagined prior to its use in the Franco-Prussian War – epoionage. In fact, the Intelligence Departments of many countries that participated in World War I made use of microfilm to smuggle important documents through enemy lines. At first the films were placed in the soles of shoes and other least suspected locations, but when more thorough searches led to discovery and accompanying dreadful consequences for the individual spy, the film was sometimes stuck to the spy's spectacles where the minute images seemed to mimic the effect of smoked glasses!

With the war ending and the need for microfilm for spying eliminated, a New York City banker, George McCarthy, sought a means of keeping track of canceled checks to avoid fraud and deception. He is credited with creating the first business application of microphotography, but interestingly, the patent for the first bank check microfilming camera, Patent No. 655,977, was received by the team of Jansen, Gardiner and Kandler as early as 1900.[51]

Mr. McCarthy understood full well the burden that storing paper checks put on banks. Checks made it possible for much speedier commercial transactions, but their downside was that they also enabled if not encouraged fraud so banks held great archives filled to capacity with canceled/used checks so they could be retrieved if necessary. When the bank itself could no longer contain these checks, offsite storage meant another rent payment, more insurance coverage, additional travel costs and one or more employees to keep the site secure.

Enter microfilm as Mr. McCarthy envisioned it. Suddenly the extra costs were slashed as the needed space was shrunk dramatically and fraud was kept at bay. In 1925 Mr. McCarthy received the patent for the "Checkograph" machine to make microphotographic copies of canceled checks that were then permanently stored in the bank. This was also a great benefit to depositors, which banks quickly began to advertise. The bank could now furnish facsimiles of disputed or lost checks whenever a customer either could not prove payment or their records were lost or stolen.[52]

Such an innovation was sure to get noticed and since it involved film it soon involved Kodak. Their interest dated from October 1926 when T.O. Babb, Kodak's West Coast manager, informed the Rochester office about the Checkograph, which enabled checks to be photographed on 16 mm film. The invention was scheduled to be displayed at a bankers' convention by Mr. McCarthy. He needed very little persuasion to include Rochester among the stops to demonstrate his machine. He subsequently parlayed a job as general manager of a new subsidiary of Kodak, Inc. that would rent, service and process the film produced by a new microfilmer based on the principles, not the design, of his Checkograph in exchange for his assigning the rights to Kodak.[53]

As opposed to Dancer, the chemist and Dagron the engineer, both of whom were "limited to experimentation with and painstaking production of micro-images," McCarthy's idea was based on simply "bringing bank checks close enough to a camera for microfilming on a big volume basis using rolls of 16mm film." Being a perceptive businessman he saw the potential for personal gain through commercial applications of his powerful idea.[54]

The era of modern microfilming thus began on May 1, 1928 in the Empire Trust Company of New York with the installation of the machine inspired by the Checkograph called the Recordak Model 1. That same day Kodak opened the first headquarters Recordak Corporation, the subsidiary McCarthy would lead, in a one room office at 350 Madison Avenue in Manhattan. Not surprisingly, the demand for the Recordak machine quickly accelerated as banks saw the advantage in having such a record of customers' transactions. Soon cities outside of New York were requiring Recordak machines. At the time there were 32,000 banks, which was an enormous challenge for the sales and service departments of the infant organization. Since

McCarthy had convinced banks of the idea of using a dual unit Recordak to make essentially two rolls of film – one for the bank and one for the Federal Reserve Bank since most checks went through there – branch locations were decided based on which cities had Federal Reserve Banks.[55]

Yet another boon for banks of check microphotography and a clear indicator of the power of microfilm to streamline processes and save time and money – characteristics that will soon become apparent with V-Mail – was the concept and accompanying reality of single posting bookkeeping. The dual-entry system had been a step forward from the days of "the hand-posting of customer transactions to bound ledgers by clerks who wore eyeshades and sat on high stools." Single-entry bookkeeping that was made possible by microfilm eliminated the necessity of the duplication of effort, of bookkeeping machinery and of the proliferation of records all the while capitalizing on the film record's accuracy! Such accuracy also acted as added protection for the bank and the depositor while saving operating costs. These improvements enabled the Recordak Corporation to successfully weather the stock market crash of 1929, since their bank customers considered its services indispensible.[56]

The next logical consumers of microphotographs were retail stores. By streamlining their process of offering bills with each individual purchase described on the receipt and instead microfilming the original sales checks and credit media and returning the records to the customers with a bill showing only the total of the purchase, billing speed was increased by as much as 300 percent. This enabled the billing workload to be reduced by 75 to 85 percent along with a reduction in the number of billing machines required. This allowed for usage of the simplest machines rather than the very expensive typewriter-accounting machines the old system required. The savings were even more irresistible to retail stores during the Great Depression years.[57]

In 1932 another arena that welcomed the microfilming of records was the insurance industry. It actually headed the list because many of the biggest companies were in the vicinity of Recordak in Rochester, New York. Given that there is a requirement to maintain records for generations instead of a single lifetime, microfilm served an immensely important purpose in being able to bring about a 98 percent saving in filing space and equipment. This allowed for millions of records to be accessible in office areas that had been kept in low-cost, but distant warehouses.[58]

1933 saw the first use of microfilming as evidence. However, it took a 1938 U.S. federal court decision to establish its usage in the courtroom. The case resulted in the finding that "Recordak" microfilm photographic copies of documents kept as "regular records" were admissible under the Federal Business Records Act (28 USC Sec 1732–33), even though not specifically provided for by that act.[59]

Microfilm's ability to accommodate larger formats – engineering drawings, newspapers and production records - opened libraries and educational institutions, industrial plants, construction, transportation and other places where enormous record files were commonplace, to the world of microfilm. In 1934 the first use of 35 mm film for preserving newspapers on microfilm took place. To accommodate the larger format of newspapers with microphotography Recordak designed a special Newspaper Microfilmer at the request of the New York Public Library as well as librarians at the *Herald Tribune* and *Philadelphia Ledger*.

In 1935, Kodak's Recordak division began filming and publishing *The New York Times* on reels of 35 mm microfilm. The American Library Association officially sanctioned this method of storing information at its annual meeting in 1936, when it officially endorsed microforms.[60]

Libraries began using microfilm to preserve deteriorating newspaper collections as well as current newspapers. Any books or newspapers that were considered at risk of decay could thus be accessed and used by more students and scholars. As we will see with V-Mail, microfilm is also a space-saving measure. In fact, in 1945, Fremont Rider who wrote a book entitled, *The Scholar and the Future of the Research Library,* estimated that research libraries were doubling in space every sixteen years! His solution was microfilm, and more specifically his invention – the microcard. This enabled items that were filmed to be removed from circulation creating additional shelf space for the rapidly expanding collections. The microcard was eventually replaced by microfiche ("a single sheet of plastic that contains several images of pages of text, whereas microfilm is a long spool of plastic film that winds and rewinds to view images of text")[61] and by the 1960s microfilming was standard operating procedure in virtually all American libraries.[62]

It took until 1937 for Recordak to introduce new microfilming machines for photographing large bound volumes and mechanical drawings in both business and industry. Another fertile territory for microfilm was our government, which kept records by the million and by the ton. The first to make widespread use of the new technology, also in 1937, were the Treasury Department and the Bureau of the Census. The former initially got involved because of its use of checks and the latter, with the enormous inundation of data every ten years, quickly appreciated miniaturization of its documents. It was the Census Bureau's need and desire to take full advantage of what microfilm offered that led to the introduction of the first microfilmer for photographing large record forms on 35mm film.[63]

Recordak sought to not only keep up with the demand for its time, space and cost saving machines, but they also sought ways to be more efficient. This resulted in 1938 in the first Recordak Junior Microfilmer. The Junior made low-cost microfilming available to small banks, more retail stores and business firms, which had hitherto been unable to justify the purchase of an unaffordable machine. The same year Recordak produced the first reversing machine for photographing the fronts and backs of documents automatically.

Finally, 1938 saw microfilm installed in hospitals for condensing and protecting patient records and case histories.[64]

Suddenly, with the commencement of a draft lottery upon the entry of the U.S. into World War II, there was a more urgent need that attracted arguably the biggest public audience in the history of Recordak microfilming history and one that touches me personally. I am now referring to the ceremonies inaugurating a Selective Service System. Another lottery, to determine who did and did not receive a draft notice during the Vietnam War, featured prominently in my first book about the Vietnam War era draft As soon as the first draft number was drawn from the fishbowl it was placed in the rectangular bed of the Recordak Junior, and all subsequent numbers were placed in the exact order of their drawing. This preserved the birthday sequence "in fixed continuity on continuous microfilm." Printing plates were then created from the microfilm record by direct projection to prevent any errors or omissions.[65]

Thanks were surely owed to George McCarthy, president of Recordak, for revolutionizing record-keeping for countless businesses, organizations and industries. In 1940 the National Association of Manufacturers awarded him the Modern Pioneer Award for contributions to the invention of the Recordak, and his pioneer work in microfilming and photographic accounting systems.[66]

Before we leave the subject of microfilm's history, it is worth pointing out that, although the internet has taken its place in many instances, there is still a source that relies on microfilm for its preservation and dissemination. I am referring to classic comic books! There are three major reasons for this phenomenon and seeming contradiction to the power of the computer that deserve mention. The first pertains to the material from which the comics were made. To save money and keep costs down to avoid passing on to their customers who were not often well off, comics were not published using high quality ink or paper. The result is they've not aged well. Enter microfiche, which ages exceedingly well decades down the road.

The second reason has to do with the very high cost of collector's items, which are what many classic comics have become. Many who would covet an original are unable to afford it and that includes libraries. Here's what's underway at Michigan State University: "The library has a comic book collection with more than 80,000 entries. But it is no longer purchasing original copies due to the 'fragility and great expense of most of these items.' Instead it's buying microfilm, which can be recreated at will."[67]

The third reason has to do with how comics have always been viewed by those who felt that they were "gutter literature". So we have a situation where the New York Public Library possesses a vast collection of comic books on microfilm, not because they sought to protect the literary form, but rather because they were once considered "unfit for a library." "The library's original policy was to microfilm the comic books and then get rid of them."[68]

I'll end the somewhat convoluted story of comic book preservation with an amusing anecdote.

> *Considering that comic books gave microfiche a little extra life, it makes sense, then, that there's a comic book about Eugene B. Power. Power is the guy who founded University Microfilms International, the company that brought microfiche to libraries around the country, in 1938. Power's company is still going strong; you may not know UMI, but if you've stepped in a library sometime in the last decade, you've most assuredly heard of ProQuest, which makes some of the most widely used library research technologies.*[69]

Is there any hope for microfilm beyond the world of comic books given its usurpation by digital technology? According to the source quoted above there is definitely a future for the medium. Here's "the secret with microfilm that will ensure its existence for hundreds of years, far longer than any hard drive or CD-ROM ever will":

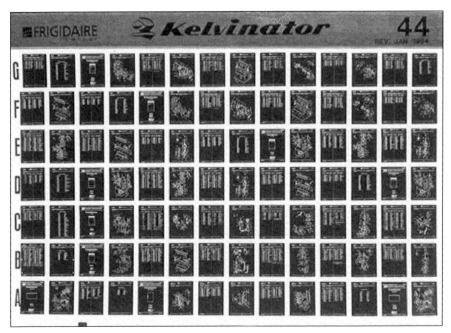

A microfiche card. (Photo: Ianaré Sévi for Lorien Technologies/CC BY-SA 2.5)

> *In a couple of hundred years, when people are trying to write the history books about our culture, they're probably going to run into a lot of 404 errors (computer glitches)...But you know what they'll be able to read crystal-clear without any issues? Microfilm microfiche – just as Paul Otlet, John Benjamin Dancer, Rene Dagron and a bunch of other experimenters might have realized back in the day.*[70]

As yet another proof of microfilm's staying power, Jan Ferrari, Director of State and Local Records Management and State Records Administrator had this to say in the Texas Record publication:

> *The reason microfilm was important in the first place was as a **preservation** tool for recorded history. A microfilm image of a newspaper or an historic map, for example, preserves that image for estimates of over 500 years, and is therefore quite stable and enduring. It is a simple, usable tool for future generations that can be used in tandem with other media. Microfilm can*

be digitized for ease of access, and digitized images can also be microfilmed. It is truly the best format to protect our history.[71]

Mr. Ferrari accompanied his analysis with this table highlighting the cost savings of microfilm vs. digitalization and paper preservation:

Since the next major development in the use of microfilm involved the specific subject of this book, V-Mail, and its forerunner, Britain's Airgraph, this is where I will temporarily pause in the discussion of microfilm's history and take it up again in the next chapter.

Table: Cost Comparisons of 1,000,000 Images Kept for 50 Years

Media	Equivalent	Unit price monthly	Annual Storage	Filming/ Scanning	Total storage for 50 years	Grand Total
Microfilm	400 reels	$.04/reel	$192.00	$42,379.50	$9,600.00	$51,979.50
Paper	500 cubic feet	$.198/cf	$1,188.00	none	$59,440.00	$59,440.00
Digital 1	50 gigabytes	$24.00/GB	$1,200.00	$87,000.00	$60,000.00	$147,000.00
Digital 2	1,000,000 images	$.00064/each	$7,680.00	$87,000.00	$384,000.00	$471,000.00

I will conclude the chapter with a comprehensive overview of the advantages and disadvantages of microfilm as seen from the point of view of libraries since that has been one of its primary uses over the past sixty years.

Advantages

The medium has numerous advantages:[72]

It enables libraries to greatly expand access to collections without putting rare, fragile, or valuable items at risk of theft or damage.

It is compact, with far smaller storage costs than paper documents. Normally 98 document size pages fit on one fiche, reducing to about 0.25 percent original material. When compared to filing paper, microforms can reduce space storage requirements by up to 95 percent.

It is cheaper to distribute than paper copy. Most microfiche services get a bulk discount on reproduction rights, and have lower reproduction and carriage costs than a comparable amount of printed paper.

It is a stable archival form when properly processed and stored. Preservation standard microfilms use the silver halide process, creating silver images in hard gelatin emulsion on a polyester base. With appropriate storage conditions, this film has a life expectancy of 500 years. Unfortunately, in tropical climates with high humidity, fungus eats the gelatin used to bind the silver halide. Thus, diazo-based systems with a lower archival life of twenty years having a polyester or epoxy surfaces are used.

Since it is analog (an actual image of the original data), it is easy to view. Unlike digital media, the format requires no software to decode the data stored thereon. It is instantly comprehensible to persons literate in the written language; the only equipment that is needed is a simple magnifying glass. This eliminates the problem of software obsolescence. It is virtually impossible to mutilate. Users cannot tear pages from or deface microforms.

It has low intrinsic value and does not attract thieves. Few heavily-used microform collections suffer any losses due to theft.

Prints from microfilm are accepted in legal proceedings as substitutes for original documents.

Disadvantages

The principal disadvantage of microforms is that the image is usually too small to read with the naked eye and requires analog or digital magnification to be read.

Reader machines used to view microform are often difficult to use, requiring users to carefully wind and rewind until they have arrived at the point where the data they are looking for is stored.

Photographic illustrations reproduce poorly in microform format, with loss of clarity and halftones. The latest electronic digital viewer/scanners can scan in gray shade, which greatly increases the quality of photographs; but the inherent bi-tonal nature of microfilm limits its ability to convey much subtlety of tone.

Reader-printers are not always available, limiting the user's ability to make copies for their own purposes. Conventional photocopy machines cannot be used.

Color microform is extremely expensive, thus discouraging most libraries supplying color films. Color photographic dyes also tend to degrade over the long term. This results in the loss of information, as color materials are usually photographed using black and white film. The lack of quality and color images in microfilm, when libraries were discarding paper originals, was a major impetus to Bill Blackbeard and other comic historians' work to rescue and maintain original paper archives of color pages from the history of newspaper comics. Many non-comics color images were not targeted by these efforts and were lost.

When stored in the highest-density drawers, it is easy to misfile a fiche, which is thereafter unavailable. As a result, some libraries store microfiche in a restricted area and retrieve it on demand. Some fiche services use lower-density drawers with labeled pockets for each card.

Like all analog media formats, microfiche is lacking in features enjoyed by users of digital media. Analog copies degrade with each generation, while some digital copies have much higher copying fidelity. Digital data can also be indexed and searched easily.

Reading microfilms on a machine for some time may cause headache and/or eyestrain.

Finally I will share two compelling quotations from Thomas Jefferson and Franklin Roosevelt on the need to preserve documents:

> *"Time and accident are committing daily havoc on the originals (of valuable historic and state papers) deposited in our public offices. The late war has undone the work of centuries in this business. The lost cannot be recovered; but let us save what remains; not by vaults and locks which fence them in from the public eye and use in consigning them to the waste of time, but by such multiplication of copies as shall place them beyond the reach of accident"*
> – Thomas Jefferson, February 18, 1791

"…because of the conditions of modern war against which none of us can guess the future, it is my hope that it is possible to build up an American public opinion in favor of what might be called the only form of insurance that will stand the test of time. "I am referring to duplication of records by modern processes like microfilm so that if in any part of the country original archives are destroyed a record of them will exist in some other place."

– Franklin D. Roosevelt, February 13, 1942

CHAPTER TWO

Airgraph: Its Origins, Uses, Successes And Challenges

In the years following the Siege of Paris and eventually World War I, the Wright Brothers invention revolutionized transportation, communication and warfare. Through the same years, as indicated in the history offered in the previous chapter, numerous improvements were made to the process of microphotography and to microfilm, such that it became a thriving business with Kodak in the forefront.

The stage was therefore set for the two inventions – airplanes and microfilm – to find expression together. Let us begin this chapter by tracing the origins of their merger beginning with some background on the state of the mail these two powerful inventions would come to assist in delivering.

Before the start of World War II, ship, train and aircraft, all operating with commensurate speed and regularity, carried mail destined for distant locales efficiently. A relatively friendly Europe protected the mail and saw to its safe and rapid delivery under the aegis of the Universal Postal Union (UPU).

Before there was a UPU, it was every country on its own having to create separate postal treaties with each nation with which it sought to exchange international mail. This had its problems, to be sure. On occasion senders would be obliged to figure out the postage for each part of a letter's journey and "potentially find mail forwarders in a third country if there was not direct delivery." It became increasingly apparent that the tangled web of bilateral agreements was becoming so complex that not only was the mail affected, but so, too, were the trade and commerce sectors of each country's economy. What were desperately needed were simplification, order and rules.[73]

The unwieldy system eventually motivated the U.S. Postmaster General, Montgomery Blair, to call for an International Postal Congress in 1863. Delegates from fifteen European and American countries were able to draw up a plan consisting of a number of general principles, but an international postal system evaded them.

Enter Henirich von Stephan, Prussian and later German Minister of Posts. Mr. von Stephan succeeded in creating a plan for an international postal union. He knew another congress was needed to endorse his proposal so he suggested that the Swiss government hold an international conference in Bern on September 15, 1874. This time representatives of twenty-two nations attended.[74]

On October 9, 1874, the day now celebrated around the world as World Post Day, the Treaty of Bern was signed. The agreement established the General Postal Union, but once the membership started rapidly expanding over the next few years, in 1878 the name was changed to what it is today, the Universal Postal Union (UPU). Thus a system with many problems bordering on dysfunction was replaced by a single postal territory with regulations for the "reciprocal exchange of letters". Barriers and frontiers were no longer a factor. Here are three of the primary rules established by the UPU:[75]

1. *There should be a uniform flat rate to mail a letter anywhere in the world.*
2. *Postal authorities should give equal treatment to foreign and domestic mail.*
3. *Each country should retain all money it has collected for international postage.*[76]

So barring war, the postal system was in place to handle the flow of mail between and among nations. Unfortunately, forty years after the UPU was established many of the world's countries were at war and the system devised by Mr. von Stephan was put to the test. What follows is the story of how twelve million letters and a million packages made it to the Front and back from England during World War I each week of the war

The General Post Office (GPO) of Great Britain already employed 250,000 men and women prior to the start of the war. It was the largest employer and the biggest economic enterprise in the country. Those post office workers who had had some military training during peacetime as reservists, the Royal Engineers (Postal Section) or REPS, as they were well-known, were brought into the Army when the war began. The Army did not have control over the efforts of these soldiers, but rather the General Post Office continued to direct their work.[77]

Mail was sorted at a Field Post Office

When the war began the REPS instantly built a gigantic wooden hut several acres in size in London's Regent's Park that was called the Home Depot. 2,500 workers, mostly women, were employed sorting the mail. Every morning the workers would learn from their bosses, who had a direct line to Whitehall, about the most up-to-date information regarding ship and battalion movements so that every mail item, sorted by military unit, could be delivered to the right place on the map.[78]

The journey to the Western Front began with a caravan of three-ton lorries conveying the mail to Folkestone or Southampton. From there the mail traveled by ship to Le Havre, Boulogne or Calais on the coast of France to the Army Postal Service (APS) depots. From there trains took over under cover of darkness. Some mail destined for locations along the route was dropped off and the rest was unloaded at railheads where REPS lorries took them to "refilling points".

From these "points" there were regimental post orderlies who sorted the mail at the roadside filling carts that would be wheeled to the front line for delivery to the soldiers. The goal was to get the letters in the hands of the soldiers with their dinner. It was said that, "No matter how tired and hungry the soldiers were, they always read the letter before eating the food."[79]

Return letters were collected and taken to field post offices. According to *Masters of the Post: The Authorized History of the Royal Mail* by Duncan Campbell-Smith, the field post office was as well-equipped as a village sub-office back in England. It was even possible for a soldier to purchase a War Savings Certificate at one of these offices precisely as the civilian population could back home. The letter was date-stamped with the specific postmark of the field office from which it originated and sent from there to the base post office for the return journey.[80]

Just as we shall find when we get to World War II, letters were censored to make sure there was no classified information that would enable an enemy who intercepted it to learn about troop or ship movements or casualties. At the outset of the war a junior officer read every letter home. It received a second reading at the Home Depot. Censorship was definitely crude and forbidden subjects were ripped out or scribbled over. Sometimes the censored words were still readable.

Despite these less than sophisticated methods and results, there were major concerns as the volume of mail increased that there could be sensitive information that, being intercepted, could result in calamitous occurrences. One indicator of the seriousness with which censorship was viewed is that when the war began there was a grand total of one person involved in censoring mail. By war's end in November 1918, 5,000 people were employed as censors.[81]

What was considered worthy of being censored? Soldiers were forbidden to write about where they were, their general condition, suspected enemy movements, action plans – "even a comment as simple as, 'Three of my friends have died'" would have caused a censor to cut out the words. Of course, the objective was to keep critical information out of the hands of the enemy. "But in reality the powers that be were more concerned that any bad news would damage morale on the home front…that if the public understood the true nature of the battlefield, support for the war would collapse."[82]

The war saw the introduction of a few new censorship options including the field postcard. These, one of which is pictured at right, gave soldiers several multiple choices, which they would cross out in order to convey their message since they were not allowed to add any other words to the card.

Yet another possible approach to corresponding with friends and family back home was the Honor Envelope. This involved having the sender sign a declaration that said that they had not included any forbidden information. Taking this form of oath resulted in having the letter be read on the home front and not by a superior in the trenches "saving the embarrassment of having their deeply personal endearments read by a censor who they knew."[83]

As was definitely true over thirty years later, the soldiers themselves carried out the greatest censorship. They had no intention of describing the horrors of war to their loved ones. They simply left out much of what they saw and did to spare their families and friends the experience of both anxiety about their safety and fear about what they confronted.

And yet…there was definitely not enough self-censoring to even imagine considering wartime correspondence between soldiers and loved ones on the home front as historically insignificant. Here's how one source described the contents of the letters:

Without doubt some soldiers did refuse to say anything that would unsettle the sleep of their wives or parents: but the correspondence of front-line soldiers, from many different armies, when read in its entirety, is extraordinarily revealing not only for what it said about the war, but also for what it tells us about how combatants remained connected psychologically and emotionally to the families they had left at home."[84]

What did make it past the censors – the French army censors eventually just sampled as little as two percent of the enormous volume of letters as the war years passed – were the soldiers' anxieties, their hopes for a better future, their love for their wives and parents and affection for their children. Souvenirs were sent home along with the detritus of battle. They begged for clean socks, parcels with more palatable food and anything to fight the epidemic of lice. They happily received letters from home delineating life's minutiae as well as the hardships that were increasing. Many wives and mothers put together packages as frequently as once a week at no small expense to men on the front lines and, even more compellingly, to prisoners-of-war. The goal above all else was to reassure the men they loved that home was awaiting their return.[85]

Here are a few examples of parts of actual letters sent from British soldiers during World War I.

Dear father… They have been teaching us bayonet fighting today and I can tell you it makes your arms ache. I think with this hard training they will either make a man of me or kill me… From your loving son, Ted"

(Edward John Poole was killed in action two months later aged eighteen.)

My own beloved wife… I do not know how to start this letter. The circumstances are different from any under which I ever wrote before… We are going over the top this afternoon and only God in Heaven knows who will come out of it alive… If I am called my regret is that I leave you and my bairns… Oh! How I love you all and as I sit here waiting I wonder what you are doing at home. I must not do that. It is hard enough sitting waiting. Goodbye, you best of women and best of wives, my beloved sweetheart… Eternal love from yours for evermore, Jim xxxxxxxx

(Company Sergeant Major James Milne survived and was later reunited with his family.)

My dear father… It is a strange feeling to me but a very real one, that every letter now that I write home to you or to the little sisters may be the last that I shall write or you read. I do not want you to think that I am depressed; indeed on the contrary I am very cheerful. But out here, in odd moments, the realization comes to me of how close death is to us… With my dear love. Pray for me. Your son, Frank

(Lance Corporal Frank Earley was killed the day after writing this letter aged nineteen)

Source: Imperial War Museums (Founded as the Imperial War Museum in 1917, the museum was intended to record the civil and military war effort and sacrifice of Britain and its Empire during the First World War. The museum has since expanded to include all conflicts in which British or Commonwealth forces have been involved since 1914.)[86]

Through this investigation of wartime correspondence during World War I, I have gained an appreciation for the degree to which "letters of lament, marked by unapologetic accounts of psychological and material misery, challenged the social convention that civilians were to endure with stoic resignation the tribulations of war." The truth, as confirmed by scholars who have read a great deal of the correspondence from soldiers from all countries involved in the Great War, is that neither the combatants nor their loved ones accepted complacently the right of the state to "censor their thoughts and render mute their grievances."[87]

Notwithstanding this caveat, the extraordinary postal system worked with incredible effectiveness, not just on the Western Front, but also delivering to ships of the Royal Navy wherever battles were being fought around the world and to soldiers wherever they were obliged to face combat. For instance in Gallipoli, more unopened letters from those killed in action were sent back from the front than letters going to soldiers fighting. These returned letters, however, did not arrive before the official telegram informing the family that their son, father or husband had died. At one point there were 30,000 unopened letters every day![88]

It is not a big leap to imagine that the postal workers who served with REPS were relieved to be handling letters and packages instead of rifles and bayonets, but the service they provided was most assuredly as important to the war effort as those fighting with weapons. "Indeed mail exchanged between soldiers and loved ones was a weapon. Those who wielded it made a huge contribution to the outcome of the war."[89]

Here is a chart of how the mail moved during World War I. It took only two days for a letter to reach the front. The journey began at the purpose-built sorting depot at Regent's Park. By the war's end, two billion letters and 114 million parcels had passed through it.[90]

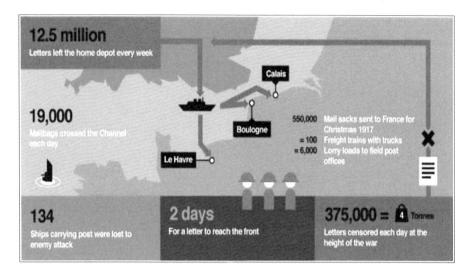

Thankfully, World War I came to an end. Many historians trace the causes of World War II to what occurred in the aftermath of the Great War, especially insofar as the peace terms were concerned, but the ensuing historical events have been written about at great length. Here we are concerned with the linkage of mail, microfilm and war, so we shall recognize that mail eventually returned to a semblance of normalcy upon the conclusion of the war and that the rules enshrined in the Treaty of Bern were once again the order of the day. This chapter began with a description of how well the mail was being distributed internationally before the war, though sadly it was to be radically disrupted again.

There is a bit of a backstory that helps explain the impetus for what followed with the commencement of another world war. This time the source is unidentified and was discovered in a file marked "Mail" at the Kodak research library in Rochester, New York by the man who inspired this book and wrote its preface, Bill Streeter. As the document refers to the other Queen Elizabeth, it must have been written after 1952 when Elizabeth II, her daughter, took the throne. Nevertheless since it is an intriguing tale, I shall share it with you here.

In the early 1930s Charles Z. Case, a Kodak employee was evidently at a luncheon with H.H. Balfour, M.P. in London. Mr. Balfour went on to become Assistant Secretary of Air and was a key player in the development of Heathrow Airport.[91] In a casual comment, Mr. Balfour mentioned the possibility and even desirability of finding a way to increase airmail volume without a concomitant increase in bulk and weight. He additionally suggested that photography might have a role to play in this effort, which inspired Mr. Case, already a major fan of microphotography, to the chagrin evidently of some of his fellow Kodak employees, to take on the challenge.

It is worth mentioning at this point, what might just be an alternate narrative (though both versions could have intersected at the time) that has the British Post Office receiving a proposal, also in 1932, to use microfilm as a means of reducing the cost of sending mail. This narrative has the proposal being rejected because the public found it unacceptable. Choose your version of this history, but either way, the possibility was shelved until...[92]

The saga continues and skips ahead to 1939, a mere two weeks before the outbreak of World War II in Europe, when Case puts on a demonstration in London for invited guests. What he proceeded to show them were copies of the preceding day's *New York Times*. What made these copies unique for his audience was that they were composed of enlargements of miniature negatives, which had arrived from the U.S. by plane. The response was one of great enthusiasm and interest, but with the imminent start of the war nothing came of the presentation and everyone instead became totally absorbed in the immediate crisis.

Only a year later, however, the mail system was in crisis and here's why:

- Ships were no longer on a regular schedule, their movements controlled instead by the Royal Navy based on security interests.

- Diversions en route made it impossible to guarantee regularity or sequence of the mail's arrival at its destination.
- The closing of Mediterranean surface mails sent via the Cape.
- Civil aircraft were required for operational needs and some were lost.
- Routes for aircraft were changed to accommodate security concerns so they were flown over Lisbon, necessitating much longer flights requiring more petrol and less payload.
- The needs of the post office were deemed secondary to the needs for space for high priority passengers and materials required to prosecute war.

In addition to all of these considerations, morale was low as soldiers were mostly hearing horrific rumors from home in the scant mail that got through along with radio reports of the bombing of cities in their homeland. In the meantime mail was accumulating at a dreadful rate adding to the difficulty of eventual delivery. Even more disturbingly, when Mussolini declared war on Britain and France in June 1940, one of his first wartime acts was to close the Suez Canal and thus access to the Mediterranean for Allied ships. As a result, mail to and from British troops serving in the Middle and Far East had to travel to England by way of the Cape of Good Hope – a detour of 12,000 miles! What this meant was that a piece of mail heading from Cairo or Bombay to London could take anywhere from three to six months. This was increasingly unacceptable as morale continued to plummet.[93]

Clearly the psychological moment had ripened so back to the Kodak story... Enough experimentation had been conducted both by Case and others, including having all correspondence between Rochester and San Francisco done by air-borne film, that all was in readiness. The experiment had yielded some undeniably impressive results that revealed both the savings in space and time. Large volumes of documents were able to be compressed into small, lightweight and easily conveyable form. The overall practicality of such a method in terms of time was unassailable.[94]

Thus operating under the assumption that the plan to use microfilmed letters aboard aircraft was eminently workable, Kodak, Imperial Airways and Pan American Airways set up a small corporation and named it Airgraphs, Ltd. with the sole purpose of handling the postal demands through airborne letters on film. Airgraph became a registered trademark of the Kodak Corporation, which was given control of the process.[95]

Re-enter Mr. Case! He was challenged by the British government to set up a system for recording letters to and from the armed services on microfilm and he was asked how long he thought it would take. His reply – three months. Needless to say, given how long it was already taking letters to make the journey, he was encouraged to get busy immediately.

And what a task it proved to be. Imagine what would be needed to get a program up and running to handle the enormous volume of letters using machines and assorted technology that had never before been needed on such a scale or for such a crucial purpose as sustaining morale in a far-flung war against dictators and fascism. What was required were photographic processing machines for military use, printers to enlarge the tiny 16mm images into four by five inch readable copies and men and women who needed to be trained to operate both the sending and receiving stations.

As if these challenges were not sufficient, two additional ones needed to be addressed as well. First, the equipment, all produced by Kodak in the U.S., had to make the ocean journey, primarily by ship. The passage was by no means secure as it had been before the start of the war, so there were big risks involved in simply getting the equipment to London let alone into the field. Second, not surprisingly, given the unprecedented nature of this program, there were no provisions in the UPU Convention for Airgraph. What this meant was that there needed to be "separate negotiations with each of the postal administrations concerned." Insult to injury? You be the judge. [96]

Despite all of these hurdles – some of which will accrue to the benefit of the U.S. version of Airgraph that is the central subject of this book, V-Mail – the system did materialize and was launched with this Airgraph letter from none other than Queen Elizabeth, the mother of the current queen. Here is that letter followed by what it looked like in its original form:

Buckingham Palace, August 1941

My dear General Auchinleck

In this first message by the new Airgraph Service to the Middle East, I wish to tell you, on behalf of all the women at home how constantly our thoughts turn to all those under your command.

I know how grievous is the separation which parts wife from husband, and mother from son, but I would assure those whose achievements have already filled us all with pride that their ex-

Queen Elizabeth taking a look at an Airgraph film.

ample is an inspiration, and I do not doubt that even greater accomplishment lies before them.

Many of them come from homes in our Dominions, and to those I send a special message of greeting. Their valour has been the admiration of the world, and to one and all I wish a speedy victory, and a safe return to their homes and those they love.

I am, Yours Very Sincerely, Elizabeth R

Here is the photograph of the actual Airgraph letter sent by the queen.[97]

The system was up and running in the three months Case had requested and the benefits were substantial. Let's cover the gains in time and space first. As indicated, sending mail to the Middle and Far East under the new reality of the war took many months to arrive if it arrived at all. In contrast, the new system enabled messages to be sent entirely by air such that ordinary surface mail to Egypt from Great Britain now took from twelve to twenty-six days compared to ordinary surface mail that took seventy-seven days and ordinary air mail that took fifty days. Mail sent to India went from taking seventy-five days using surface mail and fifty-four days using ordinary airmail to being delivered between seventeen and thirty-three days.[98]

In terms of aircraft space economy, two bags of mail weighing fifty pounds and containing 1,500 letters could be put onto a roll of film weighing a few ounces and taking up next to no space on the plane. Mail that would have necessitated the use of fifty planes could now be carried by one plane! Finally, the entire journey of the letter being sent both to and from the Front now featured solely airplane travel, which in the end was not only speedier, but safer.[99]

From the outset there were two distinct disadvantages to the Airgraph system compared to ordinary peacetime mail delivery. First and foremost there was the necessity of filming the letter twice and the time that this took as well as the machinery and training of operators this entailed. Second, as the system gained momentum and the war dragged on, the accumulation of mail owing to both volume and irregularities in flights led to unforeseen and unpreventable delays. Still, these were considered small prices to be paid given what Airgraph made possible.[100]

As regards the cost of sending an Airgraph letter, the decision was made to charge different rates for civilians and military personnel. The cost to mail a letter from home was eight pence (8d.) if being mailed to a civilian and three pence (3d.) if being mailed to a soldier. For soldiers on active duty there was no charge to write an Airgraph letter.[101]

The stamps were placed on specially designed forms, which were obtained free from the post office and on the Front lines. If the proper form was not used the letter was considered unacceptable and would not be mailed. A canceled stamp like the one below is rare, since the form was destroyed once delivery was insured. The ones that do survive are likely to have been returned for some reason or

forwarded as ordinary mail when Airgraph service was unavailable and therefore the letter was never filmed.[102]

There were additional requirements to be met in order to have one's letter sent via Airgraph. These included using black ink and writing legibly. The post office requested that the public assist them in their efforts by writing addresses in large block letters, which today's computers can type with abundant size and font choices, but the typewriters of the 1940s were ill-equipped to enlarge print. Typewriting the letter itself was permitted, but warnings were issued about the inadvisability of using a worn ribbon. "Faintly written messages, besides inconveniencing the addresses, have to be specially dealt with and thus slow up the machinery."

Here are a couple of hand-made forms:

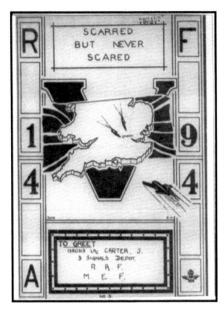
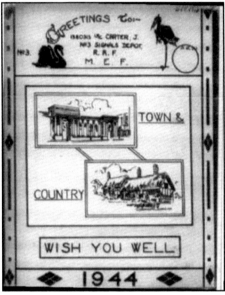

The completed forms were then photographed on "miniature film," which were conveyed all the way by air. Upon arrival photographic enlargements were made – 5⅛ by 4¼ inches, approximately one quarter the size of the original, and placed in a crude brown envelope measuring about 3¾ by 4¾ inches. These were subsequently delivered to the addressee. The original letter was retained until its delivery was assured and then the post office destroyed the letter.[103]

Perhaps you've noticed following the previous discussion of regular mail that a key step in the process is missing. It turns out that there was considerable controversy regarding several aspects of the system, but none more worrisome than that of the censor.

There was already some consternation that the letters had length limits of only 450 typed words and half as many hand-written (extra forms and postage were required for longer letters), but there were active fears that through the process of being prepared to be mailed, too many people handled the letter. When the unfolded letter (ready to be filmed, of course) arrived, first the censors had to read it. Then sorters and other processing personnel had a turn with the letter and the concern was that "the private details of one's life would become common knowledge." Provision was made for those not choosing to avail themselves of Airgraph letters, but this required providing one's own envelope plus additional postage.[104]

Here is a copy of the original instructions:[105]

```
                                                    ┌─────────────┐
                                                    │ Affix stamps│
                                                    │  value 3d.  │
                                                    │    here.    │
                                                    └─────────────┘
```

(1) The Airgraph Service is available only for messages to :—
   (a) **Personnel of His Majesty's Army and Royal Air Force serving with the Middle East Force and the East African Force** and
   (b) **Personnel of His Majesty's ships operating in the Eastern Mediterranean.** No information concerning the location of His Majesty's Ships can be furnished by the Post Office, and it must rest with the sender to decide whether or not to use the Airgraph Service. Airgraph letters addressed to His Majesty's ships which are not in a locality served by the Airgraph service will be despatched by ordinary mail.

(2) Nothing should be written on this side of the paper.

(3) The whole of the message should be written on the other side above the double line.

(4) The name and address to which the message is to be sent should be written in **large BLOCK letters** below the double line in the panel provided.
The address of Airgraph letters for the Middle East Force and the East African Force should comprise : Army or Air Force Number, Rank, Name, Squadron, Battery, Company or other section of the unit, Army or Royal Air Force Unit (including in the latter case the letters " R.A.F."), followed by the words " Middle East Force " or " East African Force," as the case may be. For example :—
   123456 Private John Brown,
      A Company, 2nd Battalion Blankshire Regt.,
         Middle East Force, or East African Force, as the case may be.

The address of Airgraph letters for His Majesty's ships operating in the Eastern Mediterranean should comprise : Rank, Name, Number if known, name of ship, followed by the words " c/o G.P.O., London, E.C.1." For example :—
   A.B. Tom Bowling, JX12345,
      H.M.S. (name of ship),
         c/o G.P.O.,
            London, E.C.1.

(5) A miniature photographic negative of the message and address will be made and sent by air mail. At the destination end a photographic print, measuring about 5 inches by 4 inches, will be made and delivered to the addressee. **It is therefore important that the message should be written very plainly and that the address should be written as large as possible in BLOCK letters.** Wherever possible, black ink should be used. The use of pencil is not recommended, but if it is used it must be a " B " pencil. Very small writing is not suitable.

(6) Stamps to the value of 3d. should be affixed in the space provided above. The completed form should then be handed in at any Post Office.
If preferred, the completed form (with stamps to the value of 3d. affixed in the space above) may be forwarded to London in an envelope which should be addressed to :—
   "Airgraph,"
      Foreign Section,
         London."
In that case it is desirable that a large envelope should be used and that the form should be folded as few times as possible.

(7) If it is desired to send more than one sheet a separate form must be completed and stamps to the value of 3d. must be affixed to each form.

(8) The original will be retained by the Post Office and eventually destroyed.

During the first twelve months of operation, ten million Airgraphs were sent back and forth between the Mother Country and the Middle East, meaning over twenty million messages. Here are some important dates in Airgraph history:

- On April 21, 1941, the first Airgraphs were sent from Egypt where they already had the equipment required to get the system up and running.
- On May 13, 1941, the first Airgraphs arrived in Great Britain from troops in the Middle East (40,000 – 50,000 letters). From posting to receipt took approximately three weeks for microfilming, transit, enlarging and delivery.
- On August 15, 1941, outward service to the Middle East began. It was confined to mail for troops. The reason for the delay in getting mail sent from England was because the first priority was getting Cairo set up to transmit mail and doing the same for London turned out to be more complicated. Nearly 1,000,000 Airgraph messages were received in Britain before the first dispatch left London. As already mentioned, the first Airgraph from the UK was sent in August by Her Majesty Queen Elizabeth (wife of King George VI and late mother of the current Queen) to General Sir Claud Auchinleck, Commander-in-Chief, Middle East Forces, conveying a formal message to inaugurate the service. It had serial number one!
- In October of 1941, Airgraph service was extended to Aden and Iraq.
- On February 17, 1942, the first Airgraph was sent to India. 539,500 letters came in the first 10 arrivals.
- On March 16, 1942, Airgraph service operated both ways to E. African troops and civilians.
- On May 21, 1942, outward civilian service to Egypt and the Middle East began.
- On August 6, 1942, the first outward service in Canada and Newfoundland was inaugurated.

Throughout Airgraph's existence all processing of films coming into the UK occurred at Kodak's factory in Wealdstone. At first, the process of inserting the letter into special envelopes was done by hand and sent on to the Post Office to receive the "Post Paid" cancellation before it could be delivered. Over time, though operations remained at Wealdstone, the manufacturer of the envelopes used by Airgraph adapted existing machinery so as to create a more comprehensive system including forming the envelope, printing the Post Paid postmark, folding and inserting the letter and finally sealing the envelope. This contributed greatly to increasing the speed of processing to 8,000 letters per hour.

Processing the Airgraph forms being sent to military personnel from the UK began with a sorting process according to service (Army, Navy, Marines, Air Force) and destination. Next came the work of the censor after which each letter received a serial number to make it possible to locate individual forms on the film roll if a problem occurred. The final step saw each letter fed individually into the machine that photographed them onto a 16mm roll of microfilm, which was developed, rolled onto a reel and boxed for the ensuing flight.

Forms heading from soldiers to their homes involved a simpler process since no sorting was required. The Post Office was equipped to sort the mail as it would pre-war with any mail that arrived. The raw forms were censored, always the fist step, and then serially numbered and photographed on Kodak's Recordak machine. The exposed film was developed and ready for shipping back to Great Britain.

Printing the Airgraphs

At their destination, the rolls of microfilm were processed for delivery to their intended recipients.

Photographing the Airgraph letter forms.

#1 Film being enlarged onto continuous rolls of photographic paper.

#2 Continuous developing tanks.

#3 Checking rolls of prints for defects.

#4 Rolls being "guillotined" into individual messages.

#5 Messages ready for inserting in envelopes.[106]

Airgraph was a significant factor in enabling people who needed to stay connected to do so and it aided in efforts to bolster morale for both soldiers and civilians. Its success is indisputable and by the end of the war processing centers overseas included Cairo, Calcutta, Algiers, Naples, Toronto, Johannesburg, Wellington, Colombo, Bombay, Nairobi, and Melbourne.[107]

Stations opened gradually as the difficulty in obtaining essential equipment and shipping challenges mitigated against speedy development of facilities. The last to become operational was in Colombo, Ceylon (Sri Lanka) in September 1944. The island was the headquarters for Southeast Asia Command (S.E.A.C.) and the number of troops was being augmented in preparation for military operations in Burma and beyond.

Since it is a subject that connects to the upcoming chapter on V-Mail, I will conclude this chapter with some words about efforts to enhance one's Airgraph letter with attachments – newspaper clippings or photographs most often. Such attachments were prohibited due to the likelihood of jamming the Recordak machine, but nonetheless some enterprising and determined folks tried them anyway and some actually got through. The thickness of the attachment was key and newspaper and magazine clippings were thin enough that if they were glued very carefully, they made it! Photographs were basically too thick, but even a few of them survived.

The desire to have photographs accompanying Airgraph letters was so popular that efforts were eventually made to provide the service. Early in 1943, Dufay-Chromex, a manufacturer of film, announced a system for adding photographs to Airgraphs. It required photographing the subject, and then printing the photo onto an Airgraph form, which took several weeks and entailed considerable labor and concomitant cost. Judging by the very few Dufay-Chromex Airgraph that have survived, this option never really caught on. As we shall soon see the same service was offered for Americans sending V-Mail letters.[108]

Over time the use of Airgraphs diminished and when cheap airmail postage to troops was offered early in 1945 the decline was intensified. The service ended in all theaters on July 31, 1945. Over the 4¼ years over 330 million Airgraphs or approximately fourteen films were processed per day. The end of the service even received treatment in Airgraph letters!

Airgraph remains a fruitful research topic for those who have an interest in postal history, for stamp collectors and for World War II historians. If you would like to learn more about the subject, see many examples of Airgraphs that are now collector's items including special holiday Airgraph forms and get a more visual experience of what was an extraordinary system, here are three worthwhile websites:

A IS FOR AIRGRAPH

Airgraph Letters (1941)

History of Airgraph from India

Here are a couple of examples of Christmas airgraphs to end this chapter. The first is indicative of the prejudice that fed the war effort on both sides. The second gives some hints as to where the letter originated, but it evidently made it past the censor:[109]

CHAPTER THREE

Victory Mail: The Patriotic Way to Mail a Letter During World War II

When the U.S. entered World War II, those in charge of mail were acutely aware of the British experience with Airgraph. Thus it was soon a foregone conclusion that the government of the latest entry into the war would adapt the system to America's needs. England, experiencing the relief of finally having a strong ally with whom to share the demands of such a war, was quick to cause Airgraphs, Ltd. to make all its patents and machines available without consideration of royalties.

Realizing that the volume of mail during wartime would far outstrip the existing system's capacity, the Post Office Department sought contingency plans even before the attack on Pearl Harbor that resulted in America's entry into the war.

In 1938 the Department began preparing for emergency mail service. Knowledge of the harmful effects of war on the mail during World War I – slow processing and delivery and lost letters – was still strong in the memories of many civilians and military personnel. On top of these logistical challenges was the awareness that disruptions to the mail had a deleterious effect on the morale of soldiers fighting and dying for their country as well as on their loved ones back home who desperately wanted news of their soldier's safety. In order to keep this from happening the Department sought a system that was both space-saving so valuable cargo could be shipped instead of endless mail sacks and ensuring of timely and efficient communication between home and the troops.

Lt. Colonel E.D. Snyder chronicles the start of V-Mail in an article in an issue of Radio News, February 1944. He served with the American Expeditionary Forces during World War I and was awarded a Purple Heart Decoration. He was then employed by the General Electric Company in the city of Rochester, New York from 1920 until he was drafted and assigned to serve in the U.S. Army as Officer in Charge of the V-Mail section, Army Pictorial Service.[110]

Snyder wrote that Colonel William Rose (later Brigadier General) was in charge of the Army postal operations. Early in 1941 he took on the responsibility to efficiently get mail to and from overseas military forces. He studied the Airgraph system and reached the conclusion that using microphotography was the best method.

At this point we learn from Lt. Colonel Snyder that the signal corps was asked to make a preliminary report. Credit goes to Major Kenneth B. Lambert (later Lt. Colonel) for taking the recommendations from the report and, after a thorough examination of all aspects of a microfilm based system, setting up the V-Mail service to be used by the United States. He submitted his report delineating how the system would work on September 20, 1941. The report included estimated costs as well as the personnel needed to make the system operational.[111]

In a letter from Colonel C. L. Williams, the Army Postal Inspector at the time, we learn that the Army Postal Service unit in the Post Office Department in 1941 consisted of himself, Colonel A.J. Kenyon and four others. He also wrote:

> *Immediately after Pearl Harbor the Army Postal Service unit in the Post Office Department was told that the President wanted to establish immediate censorship over all mail leaving or coming into the country. I took over that job and Bill Means was given the proposed V-Mail service.*[112]

We next learn from Colonel Snyder once again, that Postal Inspectors, Colonels Williams and Kenyon and Inspector Means began conducting experiments with Kodak's Recordak machine, which had been used for photographing records in the Washington, D.C. Post Office Department. Once the system appeared feasible and desirable, all rights to the Airgraph system and its technology were granted to the Army. That, thanks to Airgraph, Kodak already had processing stations in the U.S., the Southwest Pacific, New Zealand, Hawaii, the European theater and the Middle East, made the move to microfilmed mail virtually irresistible. In other places the U.S. Signal Corps opened operations.

The last minor hurdle proved to be what to call the new postal processing and delivery system. Rather than the cumbersome Army Micro Photographic Mail Service, Colonel Williams favored a name for the times and purpose of the system – short and stream-lined to epitomize the swift and compact service. Newspapers were offering stories about the appearance of the mysterious "V" for victory symbol on public buildings in Europe, so Williams came up with the idea of calling it Victory Mail and submitted it to his colleagues resulting in the following exchange as reported by Colonel Snyder:

> "Why not just V-Mail?" Captain R.K. Awtrey, U.S.N. Retired asked. "After the words we could use the notes symbolic of the V sign in Beethoven's Fifth Symphony."
>
> "Better than that," another voice suggested. "Instead of a hyphen between the V and the Mail we could have the Morse code for V!"

Thus the official name became V…-mail, which eventually was reduced to simply V-Mail.[113]

As was true for Airgraph, so too was it true for V-Mail that reproducing letters, dispatches and other documents on microfilm was an absolutely astounding space saver. By condensing the volume of mail to film rolls, thousands of tons of shipping space was preserved for war materials. The figures are mind-boggling indeed: "37 mail bags required to carry 150,000 one-page letters could be replaced by a single mail sack. The weight of that same amount of mail was reduced dramatically from 2,575 pounds to a mere 45. The blue-striped cardboard containers held V-Mail letter forms."[114]

It merits recapturing the words of Snyder from his 1944 vantage point, vis a vis the impact of the V-Mail system:

It may seem a far-fetched statement to say that V-Mail is shortening the war and thereby saving the lives of countless American soldiers, but the connection can very easily be demonstrated. All military experts agree that only offensives will bring this war to a successful conclusion for us. Beginning and maintaining of these offensives are dependent on shipping, on the staggering task of transporting millions of tons of men and material all over the face of the globe. Now all one has to do is consider the fact that each soldier overseas receives around 40 letters per month and that V-Mail saves ninety-nine per cent of the weight and space occupied by regular mail. The contribution of this function to a quicker victory becomes apparent.[115]

Can such a claim be verified? Objectively probably not, but it is a compelling argument and one that deserves to be expressed and considered in light of what was accomplished both with Airgraph and V-Mail and it certainly adds to the necessity of exposing more people to these mail processing and delivery systems since their impact on so many levels was immense.

Some further logistics need to be explored before moving onto the human impact of V-Mail. As regards the Recordak machines, about which more will be revealed in the next chapter, the demand by the armed services for them necessitated, "…discontinuing all new sales efforts, and this applies to our Customer Service work. Do not solicit any new business until further notice. It is suggested, however, that you continue constantly calling on our present customers from a good-will and service standpoint." This memorandum was sent to all Recordak sales and service people by none other than George L. McCarthy, the man who brought the Recordak to Kodak in the late 1920s.[116]

The mechanical problems of supplying sufficient Recordak machines to the military for purposes of V-Mail was significantly compounded by the loss of a large number of its employees who possessed the necessary know-how to operate the critical machines. The two schools that Kodak operated – one for using the machines and another for repairing them – immediately lost workers as many enlisted in the armed services. This internal Kodak memorandum, dated, September 3, 1942, tells the story:

> *During the past year we have lost to the armed forces approximately 90 men. If the Selective Service laws continue to operate under their present schedule, we anticipate that we will be relieved of more than 100 additional men before the end of 1943. With the problem of manpower confronting us, we naturally are turning to all sources for replacement. Here in New York we have already started training girls on service and shop work. No doubt, it will be necessary to make the same move in our other territories. Regardless of what replacements are made with girls or men over 45 years of age, we feel that our service problem is becoming more acute daily.*
>
> *Therefore this morning the only logical decision that could be made was made and all representatives are hereby notified that for an indefinite period, which no doubt will mean for the duration of the war, WE WILL NOT FILL ORDERS FOR EQUIPMENT IN CITIES OTHER THAN WHERE SERVICE REPRESENTATIVES ARE LOCATED.*[117]

It was in a *War Department Immediate Press Release* that the world learned of the imminent introduction of V-Mail. Dated March 5, 1942, it was entitled, "'V-Mail' on Motion Picture Film (microfilm) is Army's Plan to Expedite Soldier Mail." It read:

> *Seeking further to expedite delivery of mail to soldiers in the field, the War Department is perfecting a plan whereby letters will be photographed on motion picture film – to be known as V-Mail, for Victory – it was announced today. As a morale*

agency, delivery of mail to troops is second only to food in the opinion of the War department and at the present time the Army Postal Service has been coupled with subsistence, and mail goes to the troops on rations trains.

Under the set-up of the Army Postal Service, each major tactical unit, Division, Army Corps or Army has its own post office as an integral part of its headquarters. In addition there are post offices at all bases and at the present time there are more than 150 postal units in service, operated by officers and enlisted men of the Army. The Army Post Office (A.P.O.) number is highly essential to the success of V-Mail. Every soldier is informed of the A.P.O. number of the unit to which he is attached and advised to notify his family and friends of this number.[118]

The next memorandum was sent out on May 21, 1942, and was addressed to All War Department Agencies Concerned. Its source was Brigadier General William C. Rose, Assistant to the Adjutant General on behalf of the Adjutant General. It spelled out the parameters and rules governing V-Mail. At that point only the United Kingdom and Egypt were operational so letters were restricted to those locations with the memo stipulating that "the service will later be extended to other points." Only "essential communications" were at first to be accepted, with again an expectation not yet spelled out, that the service would be extended over time. The size of 8½ inches in one dimension was very clearly spelled out after which it was expressed that "the length is unimportant. Photostats, maps, drawings, blueprints or tabulations must be so arranged that they consist of separate sheets 8½ inches or less in one dimension." Folding lengthy documents was not encouraged as, "there may be difficulty in passing it through the camera and in addition the fold may obscure some essential item of information." Early on as stipulated in this memo, the promise was made that "the originals will be retained until notification has been received of the arrival of the film at its destination.

The original documents will then be returned to the sender by the Foreign Mail Room."[119]

The acceptance of the value, space-saving aspects, patriotism and speed of V-Mail was not immediate. Following its first appearance in June of 1942, families sent only 35,000 letters. However, a year later in June of 1943, several million letters were sent.[120]

The next development in the launch of V-Mail is documented in an article in the newsletter of *The American Philatelist* from July 1994, p. 608. Its author, James W. Hudson entitles his piece, FUNNY MAIL, and explains that he was the first U.S. officer trained to process Army V-Mail at Eastman Kodak Company in Rochester, NY. He writes that in the spring of 1942, Lieutenant Halvin T. Darracott, the Army's liaison with Kodak, introduced him to forty G.I.'s who he would train to microfilm letters on 16 mm film, process the film and enlarge each frame to approximately half the size of the original.

Hudson essentially described himself as the guinea pig of the commencement of V-Mail and in that role he encountered the following problems and tasks:

- lack of equipment
- minimal organization
- the need to "sell" the concept and the service to the potential users
- researching all of the documentation that preceded his arrival in Cairo on August 15, 1942
- understanding the role of the British and Kodak contracts in making the system operational
- researching and understanding the role of military security and soldier mail privacy
- interfacing with International Postal Regulations
- interfacing with the APO's
- determining postage for civilians
- determining payment to Kodak for its services[121]

The first V-Mail sent to the U.S. came from a Kodak office in London working with the British on official microfilming. The recipient was none other than President Roosevelt! The sender was Major General J.E. Chaney, the general in command of U.S. troops stationed in the British Isles. The letter was processed in Washington, D.C. as a full-size photocopy in the absence of 16mm Recordak equipment that would have produced the reduced size print that would soon be the hallmark of V-Mail. It read:

> *It is most fitting, upon the advent of the V-Mail service for the benefit of American troops in this theater, that the first message be sent to you to express the appreciation of the troops for this new service. No other single factor ties in our soldiers with the people at home so much as prompt and adequate mail service. The advent of this service is a distinct contribution towards victory.*[122]

The second V-Mail letter was also addressed to President Roosevelt and arrived the same day. It was from the U.S. Ambassador to England, John Gilbert Winant and it read:

> *I am glad to have this chance to send you a message by the first V-Mail Service to the U.S. from the British Isles. It is a fine thing to know that during the trying days of separation between those serving their country abroad and those working for Victory at home, this Service will give encouragement and allay anxiety. It marks a thoughtful recognition of family affection and the ties of friendship, which so strongly unite the American people. I know that all our soldiers here and their families and friends at home will be grateful for the Service.*[123]

On the next page is a photograph of President Roosevelt receiving one of the first two V-Mail letters sent on June 12, 1942. The service officially began three days later on June 15.

Captain Hudson ends his article about V-Mail's history by telling us that American troops called V-Mail "Funny Mail." He goes on to say, "…but they loved it. They know that their latest escapade, if the

stories passed the censor, would soon be back to the States, letting the folks back home know what they were doing. They were heroes..."[124]

The first flight of V-Mail back to Europe occurred on June 22, 1942, and the book entitled *Famous First Facts* describes this "first" as "a partial roll of film on which there were only 212 individual letters. A complete roll of film would contain 1600 letters," but the system was launched![125]

How The V-Mail System Worked

Now it is appropriate to commence a discussion of how the service actually worked. There were certainly similarities to Airgraphs, but unique features abounded and it is worth chronicling both that which overlapped and that which stood apart. To begin, here is a kit that was marketed to American soldiers heading to their assigned locale so they could write V-Mails home to loved ones. Everything needed to write home was protected in a cardboard canister. But as an on-line V-Mail source wrote in responding to this product. "I can't imagine a foot soldier lugging all this around for long." I can't help but wonder how many of these kits were actually sold and used, given the other equipment many soldiers were obliged to carry, but the idea of keeping in touch via V-Mail was certainly spreading when this was put on the market.[126]

VICTORY MAIL: THE PATRIOTIC WAY TO MAIL A LETTER

There are numerous fine resources to enable scholars, historians and descendants of World War II veterans and their families to learn about the intricacies of V-Mail, but in all my searching I only found one excellent short film that I want to be sure to pass on. It is on-line at https://www.youtube.com/watch?v=WR8cBKhgELc and in the space of about two-and-a-half minutes, in newsreel format from the time of the war, the process of sending and receiving a V-Mail letter is showcased. In addition, you can listen to military postal clerks in the field in the radio segment "4th Marine Division post office on Iwo Jima," http://s.si.edu/2idfQH9. Finally there is a fine "flipbook" computer style explanation of the workings of V-Mail furnished by the Smithsonian Institute's website: http://s.si.edu/2icUcma

I will attempt to recreate what these three resources do so compellingly, especially given the fact that two of them were present at the time V-Mail was in use. Using words and pictures I will seek to convey the ways in which this system worked so effectively to bring the soldier and his loved one into more predictable and regular contact.

The first step for the letter writer whether stateside or at war was to obtain the proper V-Mail form. Here are both sides of the form with the accompanying instructions:[127]

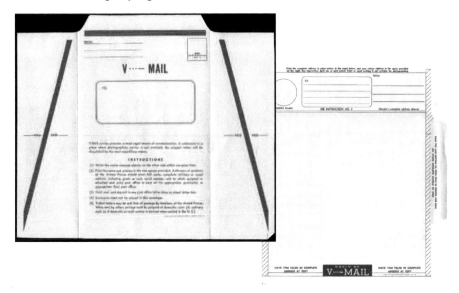

> **INSTRUCTIONS**
> (1) Write the entire message plainly on the other side within marginal lines.
> (2) Print the name and address in the two spaces provided. Addresses of members of the Armed Forces should show full name, complete military or naval address, including grade or rank, serial number, unit to which assigned or attached and army post office in care of the appropriate postmaster or appropriate fleet post office.
> (3) Fold, seal, and deposit in any post office letter drop or street letter box.
> (4) Enclosures must not be placed in this envelope.
> (5) V-Mail letters may be sent free of postage by members of the Armed Forces. When sent by others postage must be prepaid at domestic rates (3c ordinary mail, 6c if domestic air mail service is desired when mailed in the U. S.)
>
> POST OFFICE DEPARTMENT PERMIT NO. 1

The forms were initially produced and thereby supplied by the government. According to Lawrence Sherman, M.D. writing in *The United States Post Office in World War II,* "Uncle Sam picked 20 Post Offices in the entire United States and anybody who requested them could only get two forms. Requests were not long in coming."[128]

Uncle Sam gave up the exclusive printing of the blanks when millions of blanks were requested. Authorization was given to 123 private companies (a.k.a. civilian printers) once they agreed contractually to government specifications. The Government Post Office (GPO) mark used in the earlier V-Mail letter forms was replaced by a different notation in the form of a number. Permit #1 was given to the Wessel Company in Chicago, which responded by printing the forms in "astonishing quantities." Evidently the GPO did not provide identical printing plates for all printers, as there are numerous differences from company to company.[129]

A few words about the color of the blanks are in order. The original ones were black and white and the "standard model" soon became the one printed in red, but exceptions occurred. Paper was not wasted given its scarcity, so any errors or oversights that occurred were kept as "the order of the day was use all you can." One of the most striking varieties was printed and sent to India. For whatever reason they were printed in red on only one side. The A.P.O. in India sent back a simply query – what shall we do with them? The response was equally basic – finish printing them any way you can. Yellow ink was found so, yes, one side is red and the other is yellow on many unprocessed blanks left over at war's end in India.

In addition to this unusual phenomenon, some blanks were printed "locally" in yellow on rice paper! These seem to have originated in the country known until 1972 as Ceylon, which was its name in the book telling this story, but now is Sri Lanka. There are many collectors who luckily have been able to obtain copies of these rare blanks since, "though it was stipulated by the V-Mail rules that once the small version was delivered the large one was to be destroyed, a lot of these magnificent usages did survive."[130]

The public could obtain the forms in tablets and pads at local stores. As mentioned, if you went to your local post office you could obtain up to two sheets of the special stationery per day for free. Servicemen and women were able to secure the stationery at no cost and sending a letter was also free of charge. When opened the letter measured 8½ by 11 inches (plus the tab sticking out on the side to facilitate sealing the completed form).

As the pictures above indicate, V-Mail stationery functioned as a letter and envelope in one. Once the letter was finished being written a most critical step had to occur to ensure delivery – putting the recipient's and the return address at the top of the letter. This was essential because only this side of the letter was reproduced from microfilm to photographic print. One of the advertisements enticing civilians to use V-Mail included this warning (after listing all of the aspects of war that could not be controlled), "But there is one thing that you and you alone can control – the matter of the address. The most discouraging and usual cause for long delays (in delivery) is an incomplete or incorrect address."[131]

There was actually another potential problem that the letter-writer completely controlled – legibility. There were models furnished to explain what would and would not pass muster by the Post Office including this one that showed how not to type your V-Mail (outside the margins).

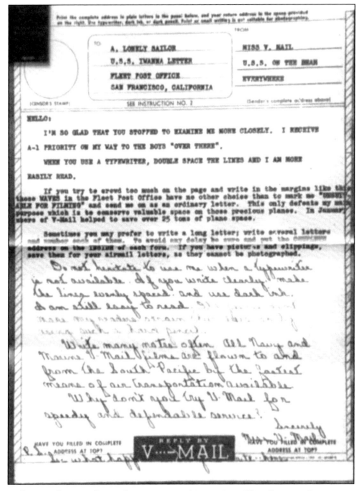

A facsimile sample with corresponding microfilm frame gives pointers on writing legible V-Mail.[132]

The final step in preparing the letter to be sent was to repeat the addresses on the opposite or "envelope" side of the sheet. This set was required to enable the mail to make the first stage of the journey from mailbox to a processing center. For anyone who might be interested in experiencing this procedure and re-connecting to what grandparents or even great-grandparents would have done, there is a website at which you can print out a blank V-Mail form to use as actual stationery. Here it is: http://postalmuseum.si.edu/VictoryMail/images/vmail.pdf. If you would like to mail the letter, be sure to use first class postage. Obviously, it will be delivered as you wrote it because the next step in the process stopped some 73 years ago…

Here is what a completed V-Mail letter looked like:[133]

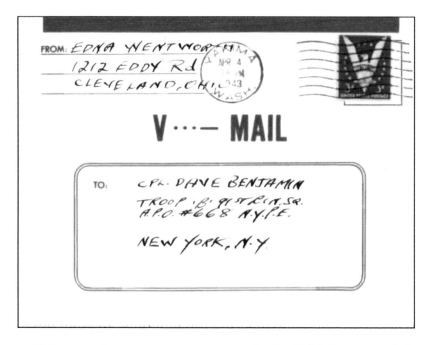

Before moving on to the next stage in the V-Mail process, it is important to get a deeper understanding of the role mail played during the war. For those overseas the importance of sending and receiving mail cannot be over-emphasized. Analogies have been made to it being almost as crucial as food and it was a form of nourishment for the psyche and spirit. The fear of loss and the need to have some sense of what was happening at home and at war gave the letters great emotional power. Were a soldier to die in battle the messages received by his loved ones would likely provide the only surviving connection between them. This was certainly true for Henry Streeter's family whose scrapbook provided many of the memories, images and returned V-Mail after he'd died and therefore never received. The life and death nature of war resulted in the expression of feelings of devotion and support on a regular basis that had hitherto only been expressed on special occasions.

The words of servicemen will serve to cement the importance of receiving letters from home. Each paragraph in the following quotation comes from a different branch of the service in this order: Army, Marine, and Navy.

> *Everybody in our outfit was feeling kind of low. Our mail came – and the next day was our second big battle. The mail made a lot of difference in the way that battle went. Everybody went into it feeling good... They had heard from home.*
>
> *You go through your battles and you don't want to think about them, except maybe that you were lucky. You want to relax, just to relax some way. Then you hear mail is coming. Maybe you hear it will come in a week and that is like tomorrow because you haven't had mail for so long. You're on pins and needles. When the sacks come, it doesn't seem as if you could wait while they sort the envelopes out.*
>
> *When the tanker pulled alongside to give us oil, they brought sacks with mail for three months. The skipper declared that afternoon a holiday for everybody except those on necessary duty – for reading mail and answering it.*[134]

The same publication that included these testimonials gave some warnings to senders of V-Mail that are certainly worth noting since the desire was to curtail the work of the censors through these directives. The title for this section of the War Department report was, "Certain restricted information should not be included":

1. Never include information concerning training, troop movements, prediction of military equipment, mention of any specific employment of that equipment.
2. Don't discuss security measures to protect plants, local utilites or transportation facilities.

3. Never include any information about the weather.
4. Don't discuss adverse conditions which affect your farm or occupation.
5. Do not include criticism of the conduct of the war. Your views may not be based on fact.
6. When sending pictures make sure no information of a military nature is included in the scene.[135]

In addition to these admonitions the War Department used this document to address several other issues including soldiers complaining that "people at home waste their money in buying and sending various presents." It quoted one soldier who summed up the frustration of his fellows when he said, "Money belts, when any guy who wore one you'd think didn't trust the men around him. And checkers – every man in our outfit got some, when the Red Cross provided all we want."[136]

Concerns about what was in store for the men "when victory is won" also received considerable attention. As one man's letter phrased it:

Will we have to wait for more than six months after duration for our discharge? Will we be stuck in an Army of Occupation? Will our jobs really be protected for us? What kind of life will we pick up when we get back? Will the fellows who are discharged first (presumably those in the States) get the best breaks? Will the Government continue to regulate employment and say where and for how much we will work? Will salaries and wages be raised for us when we get back in our old jobs to meet the increased cost of living? Will the taxes to come be so heavy that we can't return to our old standard of living?"[137]

These are indeed weighty questions that most assuredly perplexed many a serviceman and woman. The advice was plain insofar as to how to respond: "Obviously, no father, mother, wife, best girl or friend can answer all these questions. But they can write that studies already have been started to find the answers." Reassuring? Hard to say, given the hurdles those serving were facing on a daily basis, but perhaps the idea that their government was mindful of their concerns would be sufficient to allay anxiety that could only make serving more challenging.

We also learn in this War Department document about the determination of those charged with getting the letters to the troops. A general visited an Army post office that was ten to fifteen feet underground on a much-bombed airfield. "When relief was offered to the officer and twelve enlisted men stationed there, they asked to be allowed to remain at their post, because they wanted to stay up front." It might have augmented their willingness to continue working that "postal clerks in the Army carry pistols and tommy guns, for this war everyone must be able to fight."[138]

It must be pointed out that despite all of the above excitement, V-Mail received complaints as well, mostly focused on the reduced size of the print and the limits of the form in terms of number of words allowed. Of course, given that literally over a billion V-Mail letters were sent and received, these problems were not enough to diminish the impact. In addition, there were some methods that existed to overcome such problems as legibility or size of print issues. This was not a simple matter. For example, "if the V-Mail station receiving the illegible letter was in New Guinea and the V-Mail repository with the original letter was in San Francisco, how was this done?" It turns out that there was a specially printed form that enabled the requesting office to list the roll number, when the letter was sent, and the last two words of the first line of text. Few of this frequently used form have survived, but there is an example of one on the next page.

Civilians received encouragement to write to their soldiers about even the most mundane activities and to do so on a regular basis. Examples included daily routines, family news from the trite to the momentous, and local gossip all of which kept those serving connected to their homes and communities.

This surviving "Official" blank was used to request re-photographing of large (unprocess) V-Mails since they had been unsuccessfully transferred to film. Note the name of the sender and the last two words of the first line of the text as identifieersf ro each V-Mail latter that needed reprocessing. "MX 47" was the number of a particular roll, and "ZP" were the secret intitials indicating the location of the V-Mail station (in this instance, London).

War affected relationships in numerous ways. Romances adjusted to long distances as best they could – notwithstanding the archetypal Dear John or Dear Jane letter – and sweethearts and spouses separated by oceans used mail and V-Mail to maintain contact. As with all wars, couples were married on furloughs and babies were born with fathers on the battlefront. Soldiers wrote to stay connected, to sustain new and old relationships and to "fight off the loneliness and boredom of wartime separation." You shall experience the "voices of V-Mail" in Chapter Seven.[139]

As indicated in Chapter One, morale was given a major boost by letters sent and received on the battlefront and at home. Postal officials – civilian and military - supported this effort. In particular, The Office of War Information and the Advertising Council worked in tandem with commercial businesses and the community to let everyone know about V-Mail and its advantages. It was actually promoted as patriotic. Those being encouraged to avail themselves of the service were told that by using V-Mail they were aiding the war effort, which included saving cargo space and providing much-need spirit-lifting messages. To avoid fears and misconceptions, news reports including the newsreel mentioned above (https://www.youtube.com/watch?v=WR8cBKhgELc) explained in considerable detail how the process worked efficiently and successfully.

Here are actual words from one of the pitches offered by an advertisement for V-Mail by the Armed Services. One section of the "pitch" was entitled, "This is what he wants to read IN YOUR LETTERS":

- That the family is "okeh and busy"
- That the family is doing everything possible to aid the war effort
- News about his friends, especially those in the service
- Who is marrying whom
- Recollections of past events and places he used to go to
- Stories about what's gong on – and the latest gossip
- All the Sports News – particularly the dope about the home team
- News about his hobbies

All of which pieces of advice were followed by this cautionary note: "Spare him YOUR worries. He has plenty of his own. Don't mutter about civilian hardships, his are worse. Be happy and newsy. Remember, your letter might be read anywhere, under the most grueling of battle conditions. Check it over – is it fit to be read in a fox hole?" Here was additional advice from the same source specifically delineating what soldiers did and did not want to receive:

> *The kinds of letters the boys want are the cheerful, newsy ones from relatives and friends. They do not like "fan letters" written by strangers who happened to join a "Write a fighter" club. Letters from hero-worshipping youngsters or well-meaning strangers leave them cold – and those unnecessary letters may use the shipping space needed for the mail they really want.*

It wasn't just the military and post office that provided encouragement for V-Mail and letter writing. The Red Cross conducted a major campaign to promote corresponding with loved ones at home and at war. They advocated greater frequency and provided style and content tips. Civilians were encouraged to provide "positive sentiments and observations about the war and to avoid negativity and despair." In an article entitled "Sabotage Women of America," Red Cross correspondent Rosemary Ames urged women to be selective when composing their letters:

> *Men in war have neither the time nor the emotional energy to be interested in boring details about housekeeping, rationing problems and family troubles. Unfortunately, many women's minds run that way. They had better change routes for those letters are often not even read to the end. Men have told me as much. Soldiers are occupied with the fundamentals of existence. Yours, as well as theirs, only most of you are too far away from the terribleness of war and what a Nazi-dominated world could mean, to realize it. Yes, I know. It's very hard to suddenly become a psychologist and an author overnight merely because your man went away. But it's worth your while to try. For just as the right kind of letters will tighten your romances – or your bonds of affection with son, brother, or husband – so will the wrong kind loosen them."*[140]

A further fascinating article emanated from interviewing soldiers and their families. The article, "War Anxieties of Soldiers and Their

Wives" by Edward and Louise McDonagh, analyzed the influence of social forces on men and women involved in the war. Their conclusion convincingly explained in the article was "that the emotional effects of the war could outweigh physical dangers if they were left unacknowledged. Such worries were categorized into intermittent, battle, and family-related pressures." Here's what they had to say:

> *Much is being written about the G.I. and his family. And this is as it should be, for G.I.s, their wives and children, comprise approximately one-fifth of the nation's population and what this group is doing and thinking may affect America's future for many years to come. Hence, it is well for civilians to try to understand what goes on in the minds of those most closely affected by this war.*[141]

The McDonagh husband and wife team's recommendations for avoiding contributing to the traumatic effects of war included "that people be aware of war-related difficulties and write frequently to servicemen and their families to keep spirits high." Their counsel also advised husbands to remain faithful to their wives and wives to refrain from worrying their husbands over unnecessary complaints – cautioning that "gossip may travel via V-Mail to the four corners of the earth" and raise anxieties.[142]

Their final recommendations were directed to civilians, and asked them to help war-torn families:

> *Civilians can aid by trying to understand the plight of families torn apart by war, by helping to build toward goals of world peace, and by acknowledging the gratitude due all soldiers and their wives for the sacrifices of service on the war and home front.*[143]

The Navy Department also found ways to encourage V-Mail usage, especially as D-Day approached. In a Memorandum for the Press entitled Navy V-Mail Service and dated May 11, 1944, Lieutenant

James Roe wrote that more than seven million pieces of V-Mail were handled each month. The increase over the previous nine months was 560 percent. At that point, two years since its introduction by the Army and Navy, 453 million V-Mail letters had been delivered overseas. He went on to describe the results:

> It is estimated that the use of this microfilm method of correspondence has saved or made available for other vital cargo, approximately 9,000,000 pounds of cargo space. As invasion day approaches, even more cargo space on both planes and ships will be required. The lives of many American boys will depend on the prompt receipt overseas of trained personnel such as doctors and nurses, as well as blood plasma, ammunition, spare parts and hundreds of other items needed for war. The wider use of V-Mail by the public will help to speed the overseas delivery of personnel and supplies.

In some ways this last inducement to make use of V-Mail was the most desperate. Entitled "No Mail Call for Johnny" it conveys dramatically what was at stake with regard to V-Mail's role in the war. From an article entitled, "V-Mail – A Good Medicine" it reads:

> Johnny Jones was killed in action. When his possessions were sent to his mother by the Army, she found among them three first-class mail letters, full of family news. They had arrived at his base after Johnny had left. He never had a chance to read them. If Johnny's mother had used V-Mail for her letters, Johnny undoubtedly would have received them before he died. What difference does it make – Johnny's dead and the letters couldn't have saved him. Don't ask that question of Johnny's mother. She lies awake nights thinking: "Maybe Johnny was worried at not hearing from us. Maybe that worry made him careless. Perhaps if he'd had the letters and known everything was all right, he'd have been all right, too." Perhaps! But one thing is sure…V-Mail is the only mail which always travels by air… and it's the only mail on which delivery is guaranteed.[144]

This was made abundantly clear in a memo from the War Department on March 5, 1945. Entitled, "Shortage of Cargo Space Curtails Air Mail Service to Overseas Points," the memo detailed how "insufficient cargo space on aircraft available to carry the air mail load" meant that "the affixing of air mail postage on overseas mail will not guarantee the transmission of air mail by air from the U.S. to overseas destination." It went on to stipulate that, "Of all categories of overseas personal letter mail, only V-Mail will be assured of overseas dispatch by aircraft." This was ascribed in the memo to, "the large increase in the number of troops overseas and the consequent up-swing in the volume of outgoing overseas mail, which has now reached the unprecedented volume of some forty-five million individual pieces per week."[145]

Now back to the process itself. Once a stateside letter was completed, it was deposited in a nearby mailbox. From there it followed the path of any other mail, which for V-Mail, along with any other mail to military personnel meant it went to one of three APO/FPO processing centers: New York City for delivery to Europe beginning on June 15, 1942; San Francisco for delivery to Asia, starting on July 10, 1942; and Chicago for delivery to either front commencing on November 22, 1943. The cost was three cents for a V-Mail letter. It is important to remember that some V-Mail processing equipment was made available to some overseas naval units, but this was much less frequent an occurrence given machinery limitations and the constant movements of ships. Also, though there was not a sizable amount of V-Mail sent between service people, it was known to have occurred.

A word about the challenge faced due to the major holidays in November and December. Yes, V-Mail improved the speed of delivery in the ways indicated, especially all of it being air mail, but significant delays occurred due to the enormous increase in volume at this time of year. To try to avoid the delays or to make them more manageable, the Post Office Department strongly recommended that holiday mail be sent as early as September or October to have any chance of arriving in time for the holiday. Here is a letter mailed in October with tidings of joy for the upcoming holiday.

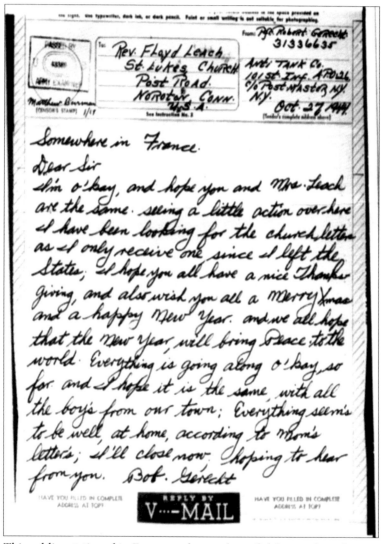

This soldier stationed in France understands mail delays and sends Christmas wished home to his boyhood pastor in October.[146]

Upon arrival at the V-Mail processing center the letter was opened to enable censors to perform their task. I found the work of Myron Fox, a past vice-president of the Military Postal History Society (a group that studies mail sent to and from soldiers) to be of particular interest and thoroughness. Mr. Fox was an expert on United States military and civilian censorship in World War I and World War II. In the interview that was conducted for the Public Broadcasting series, American Experience – War Letters, Fox offered many insights into the history and methodology of censoring mail. He begins before the Civil War and asserts that there was no "overt" censorship. His hypothesis for why this so has to do with the large number of troops who were illiterate. He also suggests that the officers were considered trustworthy and not likely to reveal any untoward information in their letters.

Since letters had to cross enemy lines during the Civil War, censorship was put into effect. The censoring that is the most well-known involved prisoner-of-war camps. The authorities at a camp like Andersonville, where many Union soldiers received well-documented horrible mistreatment, did not want outsiders to know what was occurring, so prisoner letters were censored to make sure nothing disturbing that could conceivably affect prisoner treatment in the North was sent to prisoner's families.

The first significant censorship occurred during World War I and extended into World War II. Two subjects were particularly troubling for censors – troop location and troop strength. Both of these could be of great value to the enemy. The saying, "Loose lips sink ships," was the strongly held belief in both world wars. In addition, censors were instructed to look for any morale issues for obvious reasons.

Fox goes on to point out that one of the researchers for the Military Postal History Society found over five hundred "confiscated and condemned" letters at the National Archives in College Park, Maryland. Many of these featured "graphic language dealing with sex." Evidently the same letter writer didn't get the message because numerous letters attributed to him were confiscated.

Another type of letter that was highly likely to be intercepted was one written in a foreign language. Since many soldiers were immigrants or the children of immigrants there was greater comfort in writing in their native tongue. Most censors didn't speak languages besides English so letters written in other languages were usually not delivered.

As far as the quality of the letters, in both the Revolutionary War and the Civil War, soldiers were freer to express more about their situations so that a typical letter might say, "We're outside of Fredericksburg," or "I'm in the 12th Division," information that was invariably not allowed during the two world wars. It would have been common in World War II for a soldier to write, "I can't say much or the censors will cut it out." Since soldiers couldn't write about where they were, families didn't even know if their loved one was in the Atlantic or Pacific. Instead a letter would be received on the home front with the place where the location is identified cut out of the page as well as how many troops might be in the dormitory. Efforts to circumvent the censors by writing on the inside flap of the envelope were usually discovered and thus unsuccessful.

The censoring was done by an officer in the enlisted man's unit. The chaplain or dentist would often be corralled into doing the job, which was considered unimportant by the authorities. Any soldier who didn't want his officer to read his letter – possibly as a result of mistrust of his leadership or a personality conflict – could use a "blue envelope." The letter writer would "certify that there is nothing in here that shouldn't be" and the next level of command would read it and possibly be more accepting of its contents. That was certainly the hope of the soldier.

Officers censored themselves. There was not a regular system for checking the contents of their letters. Instead spot checks by their superiors let them know they were being watched for any inappropriate information. As a result officers felt freer to say more. The reason for this is not entirely clear since it could have been because they knew better what was permitted, because they felt more confident because of their positions or because they knew their mail was censored less often – or some combination of all three factors.

Depending on the size of the section needing to be censored, those making the decisions would either confiscate the letter if it was too big or cut out the offending words if it was small enough. Sometimes they would also simply block out the words with ink. If they suspected someone of being a spy they would use special chemicals that enabled them to check for invisible writing. The offending letter would be confiscated, but no one was told since they didn't want anyone to know they had destroyed it.

Ultimately very few letters were returned. They were destroyed. The soldier whose letter met this fate was not always told what happened to his or her letter. Whether they were informed was a function of where the censor caught it and how much time there was before the troops were moving to their next destination.

In the end soldiers could not be completely sure their letters got through the censor until there was some type of confirmation from the recipient of the letter. By and large very few letters actually were destroyed so most soldiers were quite confident that what they sent would be received. They were given clear instructions regarding what could not be written, but not all followed the rules and those paid a price. That price was most likely to be a lecture from an officer. During his research Mr. Fox never found any soldier who "was severely punished for what they wrote in a letter. It wasn't considered an overt act of sabotage; it was considered carelessness."

By the time of the Korean War censorship had greatly diminished. This was attributed to the amount of time and effort that was required to go through the enormous volume of mail that soldiers sent. There were some letters that were censored during the early part of the next war in Korea. Mr. Fox and his researchers from the

Military Postal History Society believe this censorship "was an error with World War II veterans implementing World War II policy until things settled down." In addition speed of communication increased, including the delivery of mail, militating against taking the time required for the censor to read each letter. To further speed up delivery, "in the latter part of the Vietnam War, the military didn't even bother to cancel letters."[147]

As in England with their Airgraph system, with V-Mail there was deemed to be a significant need for censors to check the mail before sending it to its destination. This happened in the U.S. as well as at each field post office. Here is a letter that was written and subsequently censored on board the U.S.S. Enterprise. Its author, Ensign Arthur T. Burke, was supposed to graduate from the U.S. Naval Academy in June of 1942 only to be obliged to graduate early when he was ordered to the Pacific. He participated in the Battle of Midway and wrote the following letter to his mother and brother. You will read it as a typed transcription. The sections that have been deleted by censors in the original letter are shown as (CENSORED – plus the probable phrase):

June 8, 1942

Dearest Mom and Dan,

How's everything by you all? I am still allright. Just to let you know. Well, the Navy gave the Japs quite a licking, eh? It sure sounds like propaganda and exaggeration for (CENSORED – "Admiral Spruance"?) to say that all those Jap carriers, battleships, cruisers and transports were sunk and damaged with US losses only a damaged (CENSORED – "carrier"?) doesn't it? But it is a very conservative communiqué. The Navy is making no false or colored statements and is being very careful not to have to take any statements back, so, incredible as it may seem, it's all true and as he says himself, all the returns aren't in yet. But I still (CENSORED – "shudder"?) to think of what might very well have been had (CENSORED – "our bombers been a few"?) minutes later in their (CENSORED – "dives. How many of our carriers"?) would have been (CENSORED – "sunk"?) Oh, well!

 Say hi to everyone for me, please. How are you making out, Dan? In the Army yet? Honest folks, I am tired so I'll end this letter right here and write more when there is a chance of getting it off again.

Love,
Art

Ensign Burke was asked years later about this V-Mail letter and about the basis for the decisions regarding what to censor. He appears to be both accepting and understanding regarding what was done to his letter in 1942. Here is his response:

> The censoring rules were basically not to give information that could be used by the enemy such as ships in company, location, speed, course, plane information, damage to us or them (until such information was released officially), names that might have been codes to try to tell family unauthorized info, etc…

Once a letter had passed the censors it was sorted by where it was being sent, photographed by the microfilming equipment at a Kodak laboratory, unless it was illegible or damaged in some way or the requisite processing station was not available (about which there will be considerably more in Chapter Four) and copied onto 16mm film rolls along with 1,600 letters headed to the same location. The original V-Mail letter was then destroyed, but a copy of the microfilm was preserved in case the reel containing the letter was unable to reach its destination. (Once it was clear that the letter had made it to its recipient, the microfilm was also destroyed) The reel was placed into a mail sack and put on board the next military plane to the designated area.

VICTORY MAIL: THE PATRIOTIC WAY TO MAIL A LETTER

A sergeant compares two rolls of microfilm, together containing 3,200 letters, to an equal number of traditional missives.[148]

A clerk in the New York embarkation center compares a pouch of V-Mail to 57 sacks of mail containing an equal number of letters.[149]

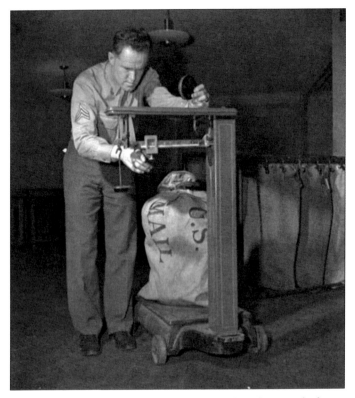

A sergeant compares the weight of a sack of mail to a reel of V-Mail.[150]

Once the plane landed safely – and not all did – APO/FPO mail centers followed troop movements. Most, but not all had V-Mail processing capability. When the mail reached its destination the microfilmed V-Mail was processed with the appropriate technology back into letter from, though as with Airgraph mail it was diminished in size, in this case to ¼ its original size – about 4¼ by 5½ inches. The new copy was folded, text facing out and placed in a special V-Mail envelope.[151]

A window displayed the address of the recipient and no additional postal markings were affixed, except in rare instances. An exceptional website for invaluable information on V-Mail's story and a source for much of what I am explaining to you, http://alphabetilately.org/V.html, had this to say about the envelope on the next page:

"Actually, I had to fake the image below – actual examples of V-Mail as received by military personnel in the field are quite rare. First, keeping personal material like this was discouraged, and second, it just wasn't practical." The opposite was true for the V-Mail letters sent back home. There were neither security nor space constraints in homes and apartments, so wives, children and parents saved V-Mail – the only tangible connection with their beloved soldiers in the field.[152]

Over the course of the war, V-Mail units were sent to all theaters where soldiers faced combat or were stationed. In the Pacific, V-Mail machine operators island-hopped along with the combat troops who "painstakingly pushed the Japanese out of the South Seas."[153]

For example, in the invasion of Tarawa, V-Mail equipment was sent in with the second wave of assault troops. Before the war ended there were V-Mail stations all over the Pacific, from Hawaii to Okinawa. It is important to mention that not only was V-Mail intended to be for purposes of letter-writing, but it was also used to send medical information to field hospitals and to reproduce essential Army and Navy booklets and information sheets.[154]

In North Africa and Europe, V-Mail personnel advanced right behind the front line troops. The stations were set up in strategic cities not far from the front lines. These cities included, Casablanca, Algiers and Oran in North Africa, Palermo and Naples in Italy, Marseilles and Paris in France and London in Great Britain. Much more on these and other stations will be presented in Chapter Five.

Throughout the war the number of military post offices overseas increased dramatically while those in the U.S. decreased. Just between 1943 and 1944 the number of APO's outside the U.S. grew from 356 to 806 stations and the number of Navy post offices increased from 2,035 to 4,869. Stations were also built in places where American servicemen manned bases. These included Alaska, India, Greenland, Iceland and several countries in South America. In many instances throughout the world, these speedy construction jobs called on V-Mail workers to build the station from the ground up. This necessitated them taking on the role of architects, carpenters, plumbers and electricians as well as being microfilm and processing experts. The results – over 1½ billion V-Mails sent and received during World War II – is a fitting testimonial to their competence and their courage.

Three dates pertaining to the conclusion of V-Mail are worth noting:

October 15, 1945:

The last V-Mail from New York Port of Embarkation is sent to General Eisenhower from Major General Kells.

November 1, 1945:

V-Mail service is discontinued.

December 31, 1947:

Free mail privilege for military personnel officially ends.

I will close this chapter with some evaluative comments about V-Mail starting with some words from General (later President) Dwight D. Eisenhower, Commander of the Allied Forces in Europe: "The inauguration of V-Mail service is a vital forward step – we, here, know and appreciate the value of the postal link that is our only connection with our parents, relatives and friends."

With these words General Eisenhower confirms what various studies conducted by World War II military authorities also showed – "that a soldier with a cheerful letter from home in his pocket was more alert then usual. Therefore he reacted faster under fire and was less likely to be injured." Quite compelling reasons for the existence and adoption of V-Mail by countless individuals!

And yet… V-Mail was not without its issues and detractors. Official Army sources put forth this summary at war's and V-Mail's end: "V-Mail service did to a large degree achieve the purposes for which it was inaugurated. It evidenced remarkable cooperation between the War, Navy, and Post Office Departments".[155]

However, the report went on to point out the "shortcomings," which included: "the problematic initial design of the letter sheet, the size of the facsimile (photographic-print letter) that was difficult to read, as well as the lack of timely determination of facilities, responsibilities, and procedures."[156] Concern about how to send a large images legibly, for example, was featured in this version of a "Sad Sack"* cartoon.[157]

George Linn, creator of *Linn's Stamp News*, captured the dilemma of the size of the print at the time of V-Mails usage:

> ...the reduction to practically one quarter size in the delivered photo often made an attached photo almost illegible and often made the written or typed letter so difficult to read that part of the equipment of every soldier or sailor should have been a Philatelists magnifying glass.[158]

---

*Sad Sack is an American fictional comic strip and comic book character created by Sgt. George Baker during World War II. ...The title was a euphemistic shortening of the military slang "sad sack of shit," common during World War II. The phrase has come to mean "an inept person" or "inept soldier." (Source: Wikipedia)

Despite these inconveniences, V-Mail prospered. It is now time to share some of the statistics that will dramatically convey V-Mail's impact on the delivery of the mail. What follows is a chart showing the numbers of V-Mail letters sent from the U.S. and received in the U.S. between July 1942 and December 1944:

1942	Sent from U.S.	Received in U.S.
July – September	3,463,094	3,689,245
October - December	7,640, 313	3,620,752
1943		
January – March	16,086,867	8,623,354
April – June	32,416,037	19,336,327
July – September	43,979,274	13,261,611
October – December	56,986,769	66,815,826
1944		
January – March	77,121,488	73,960,180
April – June	91,317,498	91,103798
July – September	78,945,125	77,514,859
October- December	64,290,234	78,081,017
Totals	572,246,699	436,006,969

That comes to a grand total for three of the four years of V-Mail correspondence of 1,008,251,668. Clearly people were making use of this system to stay in close touch with loved ones at war and at home.[159]

In an article that appeared in the public press of May 23, 1943, Americans learned that the 100 millionth V-Mail letter was written by the Allies Supreme Commander, General Dwight D. Eisenhower and sent to General George C. Marshall, U.S. Chief of Staff. It read, "One hundred million times, soldiers of the European theater have used V-Mail to send a message home."[160]

Two years later, on November 1, 1945, the War and Navy Departments discontinued the microfilming service for V-Mail. Those letters written on the V-Mail forms were sent in their original form via airmail. This continued until March 1946 when post office supplies of the letter sheets were exhausted.

Here's one final way to describe the impact the huge number of V-Mail letters had as a function of their space-saving achievement. Perhaps you will join me in appreciating the facts speaking for themselves from a fact sheet from the Office of War Information in 1944, which demonstrated the savings to the cargo space with the following tallies:

> V-Mail has saved 4,964,286 cargo pounds since its start in June 1942. This same number of cargo pounds can take care of the following items:
>
> 1. 2,298,280 units of Army K-Rations, which weigh 2.16 pounds per case when packed for overseas shipment.
>
> 2. 496,428 Garand Automatic Rifles, which weigh 10 pounds each, packed for overseas shipment.
>
> 3. 1,323,809 units of Blood Plasma, which weigh 3-3/4 pounds per unit, packed for overseas shipment.
>
> 4. 51,711 packages of Surgical Dressings, which weigh 96 pounds per package, packed for overseas shipment. An average of 112 dressings are packed in each package, which means 5,791,666 dozen dressings could have been shipped in this space"
>
> (File E-NC-148-57/180; OWI Intelligence Digests, Office of War Information, Record Group 208; National Archives at College Park, Maryland).[161]

Three important types of V-Mails convey aspects of the process not yet discussed. First is a V-Mail form letter that was necessitated by a rather common occurrence for many service people during the war – a change of address. Troop movement on land let alone at sea was frequent, unpredictable and gave rise to the need to let loved ones stateside know about the change as this V-Mail letter demonstrates:

Then there was the fateful day when a soldier learned that he was being sent home. This necessitated the word going out via V-Mail that no more letters or packages should be sent overseas and a form was also created for that purpose:

As has been mentioned at several points throughout this and other chapters, receiving mail was key to both morale and even physical well-being, so not surprisingly V-Mail even came up with a form (opposite page) to encourage keeping track of how frequently a letter would be written and sent to one's soldier. Here's what that looked like:

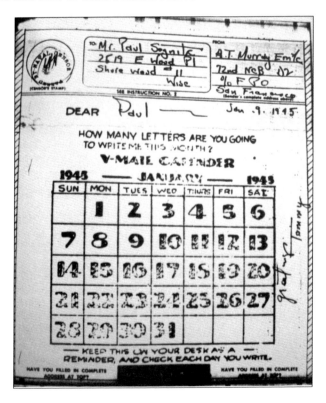

Last but not least, it is indicative of the way in which our culture takes one of its accomplishments and spreads it into other arenas. While researching the subject of V-Mail in all its various forms and influences, my good friend Bill Streeter found the photograph below affirming that the phenomenon that was microfilmed mail even entered the world of sports. Here is a V-Mail basketball team in a photo for the ages!

CHAPTER FOUR

Kodak and the Machines that Made V-Mail Work

Obviously, V-Mail relied on technology beyond conventional airmail, starting with the microfilm that began its involvement with the world of war some seventy years earlier in Europe. In this chapter the reader will be taken through the various steps in the process that enabled letters to be shipped on microfilm around the world.

What was seemingly a rather simple process involving the shrinking and blowing up of letters was actually quite involved and required both considerable people power and many machines, especially as V-Mail caught on and the war spread. First we shall re-visit the letter-writing process and then move on to the seven steps in processing V-Mail.

A soldier responding to the letter of a loved one back home.

In an article by Dr. Walter Clark on behalf of the Kodak Research Laboratories in April 1943 we are introduced to an imaginary letter-writer, Mary Jane from Columbus, Ohio. She is writing to Private John Smith, "address unknown to her, but reachable in care of the Postmaster, New York." Mary Jane obtains a V...-mail letter form free of charge at her local post office – one of the self-envelope type in standard shape, size and thickness and printed in red. She proceeds to write, in the appropriate section and legibly, John's name and approved address and in another clearly demarcated section that is dictated by the instructions on the other side, her own name and address. In the space indicated – an eight-inch square – she writes her message to John. To complete her part of the process, she writes John's address again on the outside and mails it – three cents for ordinary mail and six cents for airmail. Mary Jane has completed her task of contributing to boosting the morale of John.[162]

Now for the role played by the General Post Office, the military and most importantly, Kodak Corporation and its machines as documented in considerable detail in an article by Lt. Colonel E.D. Snyder entitled *V-Mail: The adoption of V-Mail service has reduced shipping tonnage to overseas ports.*

Step One – Sorting, Slicing and Censoring

Mary Jane's letter to John was ready for processing when it arrived at the Foreign Section of the General Post Office. The letter was sorted according to the Army Post Office designation and sent to the V-Mail Station of the New York Port of Embarkation Army Post Office. At this location it was separated into an appropriate APO bundle – if going to the European Theater of Operation (ETO) it went to New York and if going to a Far Eastern Station it went to San Francisco.

Next a slicing machine operated by a soldier opened Mary Jane's letter and stamped it with a serial number so as to be able to identify the letter

A number of V-Mail stations had machines that were designed to open envelopes and smooth the stationery in preparation for filming.

by its number if it was not delivered. There were stations where this work was done manually as shown in the photograph above.

Censoring letters even before they were microphotographed has been thoroughly explained via the interview in the previous chapter. Assuming the letter passed by the censors as almost all letters did, notwithstanding minor revisions, more often occurring with those sent from the various Fronts, it was ready for Kodak's principal contribution to both Airgraph and V-Mail – its Recordak machine.[163]

Censors reviewed correspondence from military personnel for sensitive information, clearly leaving marks and signatures. Sometimes with traditional letters they cut out larger sections but they couldn't do this with V-Mail because it would prohibit feeding the sheet into the Recordak equipment. Censors would black out questionable sections; the most common was troop location. If there was too much sensitive text, they would confiscate the letter.

Step Two – Photographing onto Microfilm

Mary Jane's letter is now at the V-Mail station in the U.S. having been sliced, censored and numbered. It is time for it to be filmed and shrunk. George McCarthy was no doubt pleased to see the application of his microfilm technology (originally invented to enable banks to reproduce an infinitude of checks to protect themselves and their customers from fraud) to the war effort via shrinking letters.

Here's how the Recordak he'd conceived of in 1928 worked in its current incarnation. As Mary Jane's letter was fed into the recorder:

>...it makes a contact that turns on the photographing lights and starts a new frame of the film reel moving synchronously with the letter to make the photograph. The letter is swept beneath a comparatively narrow strip of light, photographing a small portion of it at a time, although the movement is so rapid as to appear instantaneous.
>
>After each 100 letters a "target" letter is fed into the recorder, bearing a number. Then, if the microfilming of any of the letters is found to be faulty during inspection, the originals of the faulty letters can be easily located by the numbers on the "targets" and put through again.[164]

The target designation and number was intended to not only catch errors, but also to ensure that each group of letters would be sent to the correct overseas V-Mail station for enlarging.

There were several other devices to prevent significant errors from occurring. One was used to make certain that no letter entering the machine was overlooked. Original V-Mail letters were 4/1000 of an inch in thickness. The camera was so sensitive that it automatically stopped if paper any thicker than that was fed into the machine. This was accomplished through a Recordak innovation that provided an ingenious electronic gadget that "felt" the paper as it went through. If two letters fed through simultaneously the operator was warned and the machine was stopped.[165]

Additional safety measures dealt with the possibility of a variation in the electric current that could cause a change in exposure or the failure of the camera-driving mechanism or the film reaching the end of the roll, all of which caused warning signals and/or the machine to stop.

One last device operated if somebody accidentally opened the machine while it was operating. It immediately wound out a few feet of film to protect the part of the film that had already been exposed (the letters that had been previously photographed). Every imaginable precaution available at the time was thus taken to ensure maximum efficiency.[166]

And yet, there was one happenstance no safety device was able to prevent from causing a bit of havoc with the machinery. Under the title, "Will Bar Lipstick V-Mail – Postal Officers Set Deadline at St. Valentine's Day" the following article appeared in *The New York Times* on February 4, 1944:

> Chicago, Feb. 3, (UP) – Lipstick, "the scourge of V-Mail," will be tolerated until Valentine's Day, but after that letters to soldier sweethearts with scarlet imprints will be sent by cargo mail only.
>
> Major Kenneth H. Donnell, postal officer of the Sixth Service Command, said lipstick smears as it passes through the V-Mail photographic equipment and ruins not only the letter but also others that follow. The fast-moving automatic feeding machine must be stopped and cleaned after each "lipsticked" letter, he said. "Usually, we reject those letters and send them by regular mail," Major Donnelly said. "But we will tolerate the inconvenience they cause until Valentine's Day."[167]

Such were the vicissitudes of love and war…

The Recordak camera was able to photograph up to forty letters a minute or 2,400 an hour. A roll of 16 mm film carried some 1,600 letters in its one hundred feet.

Step Three – Developing the Roll of Microfilm

Mary Jane's letter now exists as an undeveloped photograph. We shall assume she restrained herself from adding any lipstick. Even though it was only intended to be developed to a negative for purposes of mailing, the roll of exposed film was developed on a machine similar to the one used for processing motion picture film. This involved passing continuously through a series of tanks in which developing (by time and temperature), fixing and washing occurred. Once these processes had taken place the film was dried on spools. The film that was used gave extremely high resolving power, which enabled all of the details of Mary Jane's original letter to be retained in the minute copy.[168]

Once the negative was obtained, inspectors viewed the film on projectors that caused it to be greatly enlarged. Fears that privacy would be lost were for naught because the rate of speed at which the photographs went by them was such that these folks were only able "to determine the quality of the photograph and could not begin to read the contents of the letters."[169]

The two top photos show workers – women as well as men – feeding V-Mail sheets into the Kodak Recordak machine and the one below shows original letters in comparison to the microfilmed copies.

Step Four – Packing and Shipping

Mary Jane's letter to John was now ready to make the ocean journey by plane. All that remained was to get it on board. The original letter was placed on file to be destroyed once John had safely received it where he was stationed. This was clearly the place in the process that had been the biggest motivator in that the space taken up by rolls of V-Mail was a mere fraction of the space occupied by conventional mail bags, leaving much space for crucial shipments of both soldiers and supplies.

Step Five – Enlarging the film

In order for John to be able to read the letter Mary Jane wrote without the type of equipment that was already being widely used in libraries to enlarge negatives, the microfilm needed to be enlarged. Upon its arrival the main goal was to get it back into the readable letter Mary Jane wrote. Before that could happen the film was inspected and measured on a densitometer. The purpose of this machine was to determine the density of the film. This was essential for the operators of the machinery to know the precise amount of light and stop opening for the enlarger since the density could vary from one roll of film to the next.[170]

Enlarging is a continuous process. It begins with a lens that projects an image onto the paper. From there:

> …film feeds through the enlarger at a fixed rate while below it, on the enlarging board, a roll of sensitized paper of the type known as Insurance Bromide and 4½ inches wide feeds through in an opposite direction and at a rate of speed geared proportionately to the film. They move in opposite directions because of the lens' inversion of the image. A hundred feet of film prints 825 feet of paper. The whole roll of film is thus enlarged up to a roll of paper in which the letters are printed about half the linear size of Mary Jane's letter.[171]

Step Six – Developing Paper Enlargements

The time had now arrived to get Mary Jane's letter into a reasonable facsimile of what she had written. Once the roll of paper was exposed, it was developed in a machine that is similar to the one used for the continuous development of long lengths of film. The paper

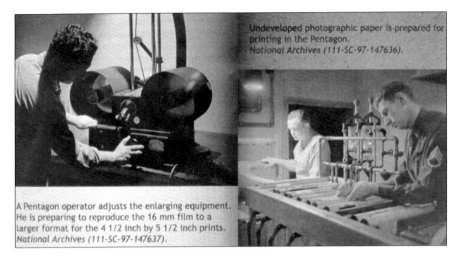

Undeveloped photographic paper is prepared for printing in the Pentagon. *National Archives (111-SC-97-147636).*

A Pentagon operator adjusts the enlarging equipment. He is preparing to reproduce the 16 mm film to a larger format for the 4 1/2 inch by 5 1/2 inch prints. *National Archives (111-SC-97-147637).*

with all of the letters on it passes continuously through developing, fixing, and washing tanks and finally over a drying drum. Yet another innovation, which was also used for processing the V-Mail negative film, enabled all of the necessary solutions to be constantly replenished to keep them at the correct levels by an automatic system.[172]

On the left in the photo above you see the enlarging equipment and on the right the photographic paper is made ready for printing.

As the paper with the letters printed on it comes from the drying end of the machine, all of the letters were in the same order in which they were photographed. After another very brief inspection, once again not affording the opportunity for even prying eyes to read the contents, any letters that are deemed illegible are re-printed in the hopes that there can be some degree of improvement so the recipient will have a likelihood of being able to read what it contains.

Step Seven – Cutting the Rolls into Individual Facsimile Letters

John is getting mighty close to hearing from Mary Jane! All that's left is her letter being individuated from the roll of enlargements, but it is at this stage of the process that we learn that this step is "the only one that has not been successfully mechanized to the fullest extent." The original intent had been for the black rectangle at the bottom of

This photo shows the manual operation of the cutting machine.

each V-Mail letter "to enable a photo-electric cell to chop the paper at the proper point," but the photo-electric cell proved to be unsatisfactory. Automatically determining the chopping point was therefore abandoned. Instead a machine was employed that was controlled manually.[173]

Step Eight – Inserting Facsimiles into Envelopes and Sending them to their Destination

At last we've arrived at the final step that leads to John reading Mary Jane's letter though the photo below is of a woman reading a letter from the Front. Once again a machine plays a critical role in making the system work. This time the developed letters are packaged and sent by special messenger to the V-Mail station. The machine that takes over is both a folding and an inserting device in one. It places each letter into a "window envelope." The address shows through the window and Mary Jane's letter along with many others is sent by airmail to its final destination.[174]

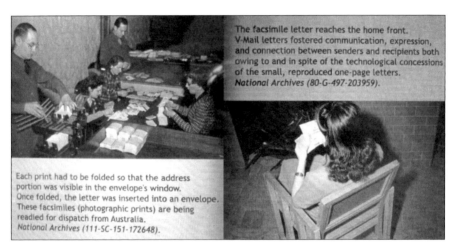

Each print had to be folded so that the address portion was visible in the envelope's window. Once folded, the letter was inserted into an envelope. These facsimiles (photographic prints) are being readied for dispatch from Australia.
National Archives (111-SC-151-172648).

The facsimile letter reaches the home front. V-Mail letters fostered communication, expression, and connection between senders and recipients both owing to and in spite of the technological concessions of the small, reproduced one-page letters.
National Archives (80-G-497-203759).

The folding and inserting tasks along with the letter being read.

Now a word about another stumbling block in the process of sending and receiving V-Mail – photographs. Initially, neither Airgraph nor V-Mail accepted letters with photos attached. In fact, the instructions at the outset were quite clear about the unsuitability of sending a photograph. Somehow by May 26, 1943, less than a year after the V-Mail system had been inaugurated photo-free, the rule was reversed and announced in an article entitled, "Photos Can Be Sent By V-Mail" in the *San Francisco Chronicle*. It began rather engagingly:

A woman operating the folding and inserting machine.

> And now, the latest picture of Junior or the girl left behind or the folks back home or the bed of radishes where the petunias once grew, is only a few days away from the fighting boys in the South Pacific or North Africa… How do we know it can be done? We did it.[175]

The author of the article, J. Campbell Bruce, pasted a small snapshot of a child onto the regular V-Mail stationery and sent it to a friend, Ken Thomson, a former paint manufacturer serving in the Seabees in the South Pacific.

> In quick V-Mail time, back came the response. "Got the photo of the little guy all right. The microfilm made him a lot littler, but it showed up clear enough. And it was the sensation around here. Everybody crowded around to see the picture sent by V-Mail. In no time at all word spread around about the V-Mail photo and before the day was out, your little guy was the best known Seabee in our base. Fellows I'd never met came around asking to see the picture. You don't know what a lift it is to get a photo of someone back home. We live for letters, but you can imagine what a thrill it is to get a picture. Send more fella!"

The article writer accurately concluded that "if one snapshot can create that stir, see what a morale builder photos by V-Mail can be?"[176] Next, however, came both the cautionary notes followed by the recommended method for ensuring that a V-Mail letter accompanied by a photograph would arrive in the hands of the recipient. First the don'ts:

> Remember, such mail, of standard form, is fed fast as human hands can shove them into a machine for micro-filming and fed flat. Anything that will jam up that rhythm won't make the V-Mail operators happy. So don't clip or pin a photo onto the V-Mail sheet. Don't slap it on with ordinary glue or paste or flour goo, so it'll dry and crinkle up and gum up the works. Don't just enclose the photo loosely in the letter and trust to luck that it won't fall – it will, or probably will be thrown out.[177]

And now the do's as recommended by a local official of the Eastman Company, "the people who operate the V-Mail for Uncle Sam":

> Cut the photo down to a suitable size, shearing off the border or unnecessary, meaningless background. Dry mount it

on the V-Mail with "photo flat," an adhesive with a rubber cement base that can be obtained at any photo supply store. It's a thin sheet tissue and will, under heat, dissolve and adhere as smoothly as the skin on your teeth. Apply this photo flat to the back of your snapshot and press it down even-edged with a hot (not scorching, though) iron. You can get enough of this stuff to send about three dozen photos for 20 cents.[178]

Following these rather elaborate instructions, Mr. Bruce added: "Or if you don't want to go to that trouble, take your photo and the V-Mail letter to a photo shop and they'll dry mount if for you. Photos by V-Mail - how about calling it 'Voto-Mail?'"[179]

Lieutenant Wes Burlingame, who was in charge of a Navy Combat Photo Unit, one of a dozen such outfits in the Navy, had this to say about the role of photographs for the men on the ships with which he was sailing:

> In censoring mail aboard ship, I've noticed that a surprisingly large percentage of all the letters I read mention photographs in one way or another. Sailors are great picture hounds and snapshots from home mean a lot when you are at sea. The sailors' letters always request snapshots, promise them or mention them in some connection.[180]

Another article entitled, "Photos By V-Mail: The Navy Says You Can – Army Says You Can't", which appeared shortly after the above piece, revealed that the two service branches took opposing positions on the issue of including photographs in V-Mail letters. It shall await further research or other means of discovering why this difference occurred, but there is one more delightful twist on the subject of including photographs with V-Mail letters that deserves inclusion. An article dated June 15, 1944, reads:

> The latest V-Mail innovation will permit sending baby's picture to daddy at the Front provided, the regulations emphasize, the child is less than one year old and was born since

daddy went overseas. The photograph must not occupy more than one-third of the correspondence space on regulation V-Mail form. It should be pasted in the upper left-hand portion to avoid creasing when the reproduction is folded, the War Department's announcer said.[181]

Once again we see the interesting combination of giving with one hand and restricting with the other as there continued to be considerable problems with folks not following directions. At the same time, there was a clear recognition of the role such photographs of very young children could play for a father preparing for battle, recovering from being wounded or even captured.

Mr. Snyder in his much referenced article, *V-Mail: The adoption of V-Mail service has reduced shipping tonnage to overseas ports,* offered a wonderful conclusion that gives voice to the way in which V-Mail brought together several key components to create a valuable communication system. It is easy to lose sight of the way in which V-Mail changed lives and offered consolation to both military personnel and those on the home front given the instantaneity of our current methods, from email to cell-phones, let alone Facebook, Instagram, Snap-chat, etc… But here's how he ended his piece:

> In the V-Mail system, photography and aviation have combined to make one of the most valuable applications of photography. It is the result of long research in photographic processes and the design of ingenious machinery, smooth working organization on the part of the Army, Navy and Post Office and the existence of overseas air-transport facilities. Along with many other developments, which have been accelerated as a result of the war, it cannot fail to be an important means of communication in time of peace.[182]

We have already seen that this last prediction proved to be untrue since the system ended soon after peace returned, but the failure of V-Mail to adapt to a post-war world and the reality that its purpose was accomplished, takes nothing away from the role it played at a pivotal time in our history.

In fact, microfilm gets credit for another major job during the war. V-Mail may have been the biggest in terms of volume of any of microfilm's achievements, but another outstanding service microfilm performed for the country was assisting in the rapid repair of damaged warships.

The Navy Department in Washington stored the blueprints of every ship in the Navy. One set of blueprints for a destroyer covered the equivalent of a quarter of an acre. The plans themselves weighed more than a ton and could easily fill a single boxcar. When a wounded ship limped into port, the Navy shipped "literally carloads of blueprints to that port to assure accurate repair work. After the desolation of Pearl Harbor, the Navy could not afford to wait endlessly for blueprints."[183]

Voila, microfilm! In Washington the blueprints were photographed on 35mm microfilm, using the reliable Micro-File Recordak. A single officer was thus able to carry the plans by hand and by air if necessary to the appropriate port. There was even an example of the procedure happening so fast that the repair sections had been fabricated and were ready to install when the damaged vessel arrived at the West Coast port. The Navy saved twenty million dollars for American taxpayers by recording engineering drawings on microfilm![184]

Microfilm also played an important role in a manner in which its usage had been contemplated a hundred years earlier in its formative stage. It served the Armed Forces in a variety of intelligence work. Maps, documents and other essential information were microfilmed in Washington and carried by plane to military outposts around the world. The speed and safety – as was equally true about V-Mail – were not possible in any other way, except by radio, which left open the possibility of decoding by the enemy. The way it worked in one instance was that portable Micro-File Recordaks on Guam filmed Japanese charts, documents and code books and then sent them on to Washington and Pearl Harbor.

What this gave rise to was another invention by Recordak– a portable projector that weighed less than twelve pounds. It could be transported with the film and used on any wall surface. It was designed to accommodate both 16mm and 35mm film, so it could be put to use for V-Mail projection of letters as well.

The Navy Department conducted one last use of microfilm upon the conclusion of the war. It was commissioned to roam the devastated countries in Europe and, using a portable Micro-file Recordak, to photograph all evidence for the upcoming trials in Nuremburg. They succeeded in uncovering hundreds of thousands of German scientific and technical documents that aided massively in the prosecution of the war crimes cases.

With such a realization of the time-limited value of V-Mail, not to mention microfilm's achievements, this chapter's mission is complete and it is on to an examination of the V-Mail stations that made all of the above steps possible.

CHAPTER FIVE

V-Mail Stations Around the World

An article featuring stories about the location and construction of V-Mail stations begins by letting the reader know that, "The Army and Navy give V-Mail the same shipping priority as they give medicine." The article proceeds to describe in words and pictures the enormous efforts put forth to provide processing centers, called stations, globally, yet another recognition of the enormous importance mail had for servicemen and women and their loved ones back home.[185]

There were four V-Mail processing stations in the U.S. The largest was in New York City and was also the first to open on June 22, 1942. It was known as the Official Photomail Station and it handled mail with APO's for the European Theater. San Francisco's station was the second to open on July 10, 1942, and dealt with V-Mail for the Pacific area. Chicago's station opened third on November 15, 1943, and handled V-Mail for mid-western areas with a staff of over 420 postal workers.

Women were essential to the smooth operation of the V-Mail system. In the publication from the Office of War Information entitled "Fortnightly Budget: For Wartime Editors of Women's Pages," February 5, 1944, one learns that:

> Every two weeks about 30 WAVES (Women Accepted for Volunteer Emergency Service) are graduated from the Navy's Postal Service Training School in Sampson, NY. The first class was graduated on Nov. 15, 1943. Today 160 WAVES are serv-

Official U.S. Navy photo of Specialist (mail) Third Class Charlotte Bennighof of Minneapolis, MN working at the V-Mail folding machine in the Washington, D.C. Navy post office.

ing at Navy post offices, releasing men for overseas duty. The course at Sampson lasts 8 weeks and covers all phases of Navy postal service, including V-Mail.[186]

In New York all V-Mail was handled from 385 Madison Avenue. The letters from eighteen states were processed there and the Office handled mail to and from Europe, China, Burma, India, Egypt, the Persian Gulf, Iceland and Africa all of which had processing stations set up to receive mail and get it to the recipients. The New York office employed eight hundred members of the WACS (Women's Army Corps). The San Francisco office handled mail to and from New Zealand where it was microfilmed and sent to the South Pacific, New Caledonia, Fiji, Honolulu and Australia. The Chicago processing station met the needs of approximately twenty mid-western states with a staff of over four hundred. It was opened to not only deal with the enormous volume of mail, but to also "result in saving about thirty hours of time required for the average V-letter to reach its destination," Major Kenneth Donnelly told *The New York Times* on November 7, 1943.[187]

Before addressing the creation of postal stations around the world, which were set up in disputed territory even while offensive operations were in progress, a word about the safety and reliability of V-Mail from the point of view of the planes transporting it. It turned out that even though planes were shot down with some frequency early in the war, loss of mail was almost zero.

One plane was famously shot down over Lisbon, Portugal resulting in the tragic loss of the British actor, Leslie Howard, of "Gone with the Wind" fame among many roles he performed during his distinguished career. He was a victim of shell shock during World War I and his activities during World War II included both acting and filmmaking. He took an active role in anti-German propaganda and was rumored to have been involved with British and/or Allied Intelligence, which sparked some conspiracy theories when he died in 1943 at the hands of the German Luftwaffe.[188]

186 rolls of microfilm containing some 336,800 letters went down with the plane. The letters were promptly re-filmed and on their way from New York the very next day![189]

Also worth mentioning, since it was indicative of the high volume of mail being processed during the war, was the opening of a second New York Port of Embarkation Post Office in Long Island City, Queens in September 1944. *The New York Times* story about the grand opening at which Postmaster General Frank C. Walker was presiding highlights that, "The Post Office Department is faced with the tremendous task of handling 80,000,000 to 90,000,000 Christmas packages going to American fighting forces overseas." The new installation was designed to relieve pressure on the Army's existing facilities in New York. Its purpose was to handle parcels, newspapers and magazines intended for shipping to foreign battlefronts.

The manpower needs were truly extraordinary, especially during the holiday season when 4,000 soldiers and 10,000 civilians sorted and shipped parcels just from the new branch. The facility in Long Island City was said to be the world's largest building of concrete and cinder block construction. It covered 14½ acres and was built in a mere 3 months, a record for speed since such a building would normally have taken at least a year and a half to complete.[190]

These construction records give us insight into what was occurring with postal stations all over the planet. What follows is an overview of developments as stations needed to be built very quickly, machinery installed and workers provided to get the mail to and from servicemen in numerous theaters of war. We'll start with North Africa.

A September, 1943 article from Kodak's own in-house publication known as *Kodakery* is entitled "African Belles Aid V-Mailers". It describes, "a bevy of North African belles is helping three former Kodak employees man a Navy V-Mail station somewhere in Africa." No doubt the location was confidential since anyone could read the newsletter and convey such information to the enemy. David Mead of Washington Processing, one of the Kodak employees, commented, "We believe we have the best V-Mail setup in the Navy. Because of the manpower shortage here we have a bevy of local belles to help us. To our surprise, they are very satisfactory, but I won't try to name them. I can't pronounce half their names, let alone spell them."[191]

The article goes on to describe the workload. "By working seven days a week and between twelve and sixteen hours a day, the station is handling six times the amount of work that was estimated for each station," a definitive example of what the armed forces were up against in maintaining delivery to uplift morale. Mr. Mead then described the challenges his unit faced: "We have to mix chemicals in salt water and wash film and prints in it also. Although we have three nice darkrooms, we lack proper ventilation for all of them. After a few hours in one you feel washed up for the day." He continued his description by telling how they cooled developer by putting the can on ice and waiting until it reached 67 degrees. "Then we grab it and run like hell for the darkroom so we can use it before its temperature gets back up to 100. I'd sure like to be back in a nice air-cooled darkroom on one of those eight-hour tricks," he concluded.[192]

From North Africa we travel to Hawaii, which had special circumstances of its own with which to contend. Since the attack on Pearl Harbor on December 7, 1941, the islands had been under total blackout after 8:00 p.m. In an article in *Kodakery* dated September 16, 1943, Les Goda, Jr., manager of the V-Mail Department of Kodak, Hawaii explained, "that the blackout has really affected the lives of the employees (working on processing V-Mail) in Hawaii." The curfew was eventually lifted to 10:00 p.m., but "those who work at night must obtain special passes, and since all buses stop running at 9:00 p.m. the shift workers who leave the lab about midnight must be provided with transportation by the lab."[193]

The Casablanca V-Mail station is described as it evolved beginning on March 7, 1943, in an article entitled, "V-Mail – A Good Medicine." It began as a group of pup tents pitched in a three-hundred square-foot barley field.

The facility remained in this state for four days largely because experienced engineers were needed nearer the battlefronts and were thus unavailable even though V-Mail was of the highest priority. As a result the V-Mail servicemen who had been recruited, often from Kodak Corporation, became their own engineers and construction crew. Here they are laying the foundation for the V-Mail laboratory a mere four days after the preceding photograph was taken.[194]

In an effort to speed construction to make the mail service available as soon as humanly possible, native Moroccans assisted, including laying a water pipe line as depicted in the photograph on the next page, taken on April 12, 1943. Three days later the first V-Mail letters sent to American soldiers in North Africa were moving through the still make-shift station even though the generator was not installed. Instead wires were attached to an Army truck to supply electricity needed to process V-Mail.[195]

The article goes on to describe additional difficulties and how the lemons got turned into lemonade as follows:

Transportation of supplies was extremely difficult in those days. Some were lost in transit. Some of that which arrived safely had to be diverted to combat areas. But the V-Mail men weren't stumped. Packing cases, which had brought equipment were ripped up and used in the construction of necessary buildings. On May 4, 1943, these "packing box buildings" looked like this:[196]

Then there was the need for water – for both drinking and for processing V-Mail. In Casablanca a huge water tower had to be built and hoisted onto a roof. Of course, darkrooms and chemical mix rooms were required, but so precious was water that the mix room had to have a device on its outside for cooling waste water to use again. The entire process including the water tower was completed in less than three months so that by August 1, 1943, V-Mail was "going through like a breeze."[197]

The first V-Mail Station in the Persian Gulf Service Command was written about in the November 9, 1943 issue of *Kodakery*. It opened on October 22, 1943, with Lieutenant Ralph Kohl of Kodak Park's Research Lab in charge and with eight Kodak servicemen working with him. In addition to being a V-Mail processing center, the command was the Army's supply line to Russia.[198]

What follows exemplifies the challenges faced not just by the men working in extreme conditions, but also by the equipment they were using to make the V-Mail system efficient. Lieutenant Kohl wrote that:

> …the supplies had crossed the equator twice and laid out in the sun for many days, the sun temperature ranging from 150 to 180 degrees. I had numerous tests made and the results were as follows – a slight edge fog on the film, but still usable, the same on the Insurance bromide paper. So you see the quality of good old Kodak products is exceptional.[199]

Hawaii, Casablanca and the Persian Gulf V-Mail stations had their share of struggles for a variety of unique reasons, but what about the station that accompanied the assault troops as they invaded Namur, Kwajalein Atoll in the Marshall Islands? Under the headline "Mail Unit Lands Quickly" in a February 15, 1944 article, we learn that, "A United States Post Office was set up and doing business on this islet yesterday less than 24 hours after Marines had landed. Today 5000 letters were packed aboard a Navy plane that left for the States." The necessary equipment landed with the invading force and stamps were for sale even before the island was secured. "Apparatus for sending V-Mail, 20,000 V-Mail forms and $1000 worth of stamps (for conventional air-mail letters) were included in the unit's stock."

Captain Emmett Harding of Hempstead, New York said, "This was the first time a post office unit had landed with assault troops."[200]

It wasn't proximity to the enemy that provided the challenge for the V-Mail station on Iceland. Instead it was the elements. Acknowledging that it wasn't the ice or snow, but rather "we do have wind. A very disconcerting wind... It can blow from all points of the compass at once. Yet we soon learned that a few minutes' wait will see a storm changed into a clear, sunny day." The midnight sun of summer was a bit of a stumbling block for those stationed on Iceland, but as a *Kodakery* article from June 20, 1944 acknowledges, "... as time passed we became used to the sudden changes in the weather and have never found the going really rough. The boys in the tropics with insects, snakes, heat and humidity are much worse off."[201]

Next in the article about conditions and considerations in Iceland came a list of difficulties worth contemplating. "As for our work, we have had our difficulties in processing V-Mail: lack of adequate water, cold developing rooms and solutions, intermittent electric power – even with our own generators." Credited with overcoming these obstacles it is, "Yankee ingenuity that comes to the fore." Sounding a bit like a Rube Goldberg system here's how they applied their version of Yankee ingenuity to V-Mail processing:

> ...a motor here, some pipe and fittings there, a bit of wire and a few odds and ends, and you've a time- and energy-saving system that takes water out of a jeep-pulled water trailer, pumps it into an overhead storage tank and then into our version of a hot-water circulator. The whole unit may look like something out of this world to a plumber, but to us it adds up to more dependable developing.[202]

With this acknowledgement of the unconventionality of the make-shift machines and processes, the article's author continues to indicate what would not appear standard operating procedure to a Kodak man:

At first glance the equipment we use looks standard, but if the Kodak men who designed and built it were to take over this station they would need an illustrated book of local directions. However, at all works and facilitates prompt handling of mail. Also, it has been a contributing factor in our reputation as the field station which consistently turns out the best V-Mail...Speed in handling the mail means a boost in morale for the men holding the "Rock." And the quality of work we have been able to produce under adverse conditions gives us real satisfaction.[203]

The article ends with a strongly worded appreciation and an enthusiastic endorsement of the war effort: "And to you folks back home, thanks, sincerely, for having done so well by us. The thought, the time, the labor and the money that Kodak and Kodak employees have given so unstintingly to support the armed forces with the best possible tools of war are appreciated. Your efforts, complementing those of the men in service, can only lead to victory." Again high marks and much praise are richly deserved for the role of V-Mail and those who participate in it as both workers and letter-writers.[204]

On South Pacific islands, V-Mail stations were located in Quonset huts measuring forty-eight feet long by twenty feet wide. In the photo below, V-Mail operators are removing processed letters from the developing bath in one of the huts and putting them through the drier before inserting them in their envelopes for delivery to the recipient. As seen in the photograph on the preceding page, the V-Mail workers themselves made the chairs in the picture from packing cases.[205]

It turns out the Navy V-Mail stations in the South Pacific had their own newspaper to enable them to keep in close contact. Photographer's Mate, second-class, Hank Doell, formerly with New York Recordak, was the founder of "V-Pac News." The microfilm of the paper was forwarded to the Navy and Marine stations where it was blown up and distributed to each V-Mailer at each station.[206]

The first issue of "V-Pac News" included a personal section, which featured the news that Doell was the proud owner of a collection of

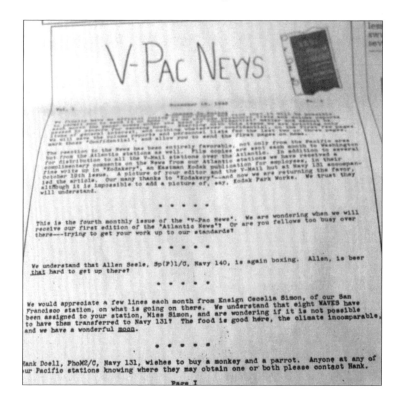

seventeen lovebirds and four Bengalese sparrows all kept in a large cage. It mentioned that he was soon to become a "papa" since seven lovebird eggs were in the hatching process. Here's a copy of the first page of the fourth "V-Pac News."

Speaking of newspapers, an attempt was made during the course of the V-Mail system's existence, to print special editions of many different newspapers from around the U.S. These would be photographed on microfilm and forwarded to local residents of the city or town whose newspaper would then be available to those serving abroad. This service was begun in 1943, but was exceedingly short-lived when it became abundantly clear that "the idea's popularity would swamp V-Mail facilities."[207]

Another community effort involved the Boy Scouts. In the February 1945 edition of "The Local Council Exchange," which was a Boy Scout publication, the suggestion was made that units participate in "The V-Mail Scout Plan." This entailed having scouts in, for instance, Newtonville, Massachusetts being given pre-addressed V-Mail forms to write to local boys serving overseas. The results were recognition for those scouts who participated and a personal interest in the war effort. The program proved quite popular. Secondary gains included an increase in those interested in scouting.[208]

A *Kodakery* issue from December 12, 1944, describes the station "somewhere in the China-Burma-India area where up to 500,000 V-Mail letters were handled weekly." It refers back to the first V-Mail film roll sent out over a year earlier on November 27, 1943, from the "hastily erected and incomplete installation." Thereafter more modern machinery was added to streamline the rapid handling of incoming and outgoing V-Mail letters. These key improvements included: paper processing machines that developed, fixed and dried 68,000 letters every twenty-four hours; an enlarger turning out 3,600 letters every hour; automatic film developing machines capable of handling 8,200 letters per hour and an inserting and sealing machine that stuffs 3,600 V-Mail letters every hour.[209]

As anyone who knows this author is aware, synchronicity has a very special appeal so please indulge me as I present you with my favorite coincidence from the research for this book. In the October 10, 1944 issue of *Kodakery* there is a photo of Elmer Dengler with the following caption beneath:

> Mrs. Rose Dengler was pleasantly surprised recently when she went to a movie and saw her son, Elmer, on the screen as a member of a V-Mail unit in Italy. Elmer is a technician, fourth grade who went into the service from the Cine-Kodak Processing Department, Kodak Park. After the show, Mrs. Dengler told her story to the theater manager who had one of the frames cut out showing Elmer. And here he is:[210]

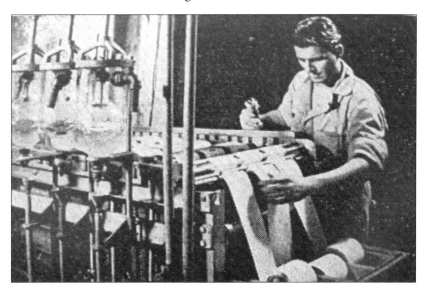

Perhaps nowhere were conditions as extreme as they were when the station was established on Iwo in the South Pacific. The *Kodakery* issue of February 27, 1945, begins the article about the V-Mail station with these words: "Life is still a little on the rough side here and lacking a few niceties, but at least there is a post office now. The establishment was set up this morning 1,500 yards (less than a mile!) behind the front lines by the Fourth Division postmaster, Captain

Emmet E. Harding." Equipped to handle 100,000 outgoing pieces of V-Mail daily and any amount of incoming mail, the station was located by Harding in an abandoned cistern, partially blown up by American guns. The V-Mail laboratory was set up and immediately ready for business, handling mail from all Marine divisions. Within a very few hours, "a big batch of mail is ready for the first mail plane scheduled to leave for Saipan tonight."[211]

These and many other stations made the system work and great credit is due to those who constructed and manned these postal outposts. Were it not for their unsung yet heroic efforts it is difficult to imagine how American soldiers would have sustained their morale.

CHAPTER SIX

How Advertising and Artwork Popularized V-Mail and Made It More Appealing to Receive

V-Mail served many purposes and it became the work of the government and private businesses to make it the preferred means of corresponding for Americans during World War II. This was done through a variety of methods. By the start of the war, since advertising was established as a key feature of the world of commerce in which people were encouraged to purchase goods and services, not surprisingly it had a significant role to play. This chapter will focus on the business world's contribution to incentivizing V-Mail letter writing as well as to the contributions of artists to make the appeals more attractive and therefore more likely to work.

Letters from home were cherished. Advertisers compared the receipt of a letter from a loved one to "a five-minute furlough" as in the ad below from a magazine produced by Martin Aircraft. The advertisement was encouraging V-Mail use going so far as to suggest, "Why not read this magazine later and write a V-Mail letter now." It didn't stop there. It went on to sing the praises of the Martin Mars plane. By focusing in on the way their plane "boosts morale" by delivering V-Mail letters – if only carrying such letters it could bring 260 million – the company is promoting the use of V-Mail because "it gets there quicker and saves space for vital supplies that help speed Victory."

The Next Best Thing to a Leave ··· is a LETTER

★ Home is heaven to men overseas. And a letter is a five-minute furlough at home. So however busy you are, find time to write that man in the service. When you write, remember these 3 rules: 1. Short, frequent letters are better than occasional long ones. 2. Write cheerful newsy letters about familiar places and faces. 3. Use V-Mail, because V-Mail gets there quicker, saves space for vital supplies that help speed Victory. Why not read this magazine later and write a V-Mail letter now?

HOW THE MARTIN MARS BOOSTS MORALE
Mighty morale-booster is the Martin Mars. Each trip she carries thousands of letters ... and if loaded only with V-Mail, she could carry the unbelievable total of 260,000,000 letters! Looking ahead, this great capacity of Mars-type planes will mean greater payloads, lowered costs, for overocean airlines. Already designed, commercial cargo and passenger versions of the Mars await only Victory to become reality. So tomorrow, for speed, safety, comfort and economy, plan to take trips or ship goods via Martin Mars!

The Glenn L. Martin Co., Baltimore 3, Md.
Glenn L. Martin-Nebraska Company—Omaha

V-MAIL VIA MARS! Making as many as 14 trips each month between California and Honolulu, the huge Martin Mars speeds mail, supplies and priority passengers to the Pacific. A number of 82-ton sisterships will soon join her to serve our lengthening battle lines.

Not surprisingly, companies like Martin Aircraft saw the benefit to the war effort and to their bottom line that would accrue from both endorsing V-Mail as a means of ensuring more dependable mail delivery and going to considerable lengths to point out their pivotal role in bringing about a quicker end to the war. The article ends with this ringing endorsement of their product:

Looking ahead this great capacity of Mars-type planes will mean greater payloads, lowered costs, for over-ocean airlines. Already designed, commercial cargo and passenger versions of the Mars await only Victory to become reality. So tomorrow, for speed, safety, comfort and economy, plan to take trips or ship goods via Martin Mars.[212]

Another pitch from advertisers involved selling the potential V-Mail user on it being preferable to traditional airmail letters. In this ad paid for by Knott Hotels (at the bottom of the ad the hotel chain has made sure the reader – of the fine print, at least – sees that of the 5000 employees of the chain "over 1200 are serving their country in the Armed Forces") V-Mail is touted as being FAST, SURE, CONFIDENTIAL ("no unauthorized third person reads it" – you now know that authorized third persons most certainly did) and helpful to the war effort since it saves vital cargo space. The advertiser takes it one step further: "A letter each day, via V-Mail is preferable to one lengthy letter a week sent ordinary mail which is often delayed or lost in transit."

Aluminum Company of America in one of its advertisements linked up the purchase of war bonds to the writing of V-Mail letters. The soldier asking "Can you send me this kind of letter?" is holding a mock letter with the date of June 10, 1943 – the war still raging on all fronts – all about the ways in which those at home are "just plain lucky" compared to the "boys who are risking their lives for me. I can never re-pay them, but at least I can try," by investing in war bonds that support the war effort.

The premise of the ad and the letter it depicts are stated in the first line: "Uncle Sam wants our fighting men to know what the folks back home are doing to win the War." It goes on to describe an America so very different from the country of the twenty-first century wars in which most of its citizens are not indelibly touched by the casualties – physical, emotional, psychological and spiritual of war. "Lives there an American who does not have someone close to him or her in the Army, Navy, Marines, Coast Guard or Merchant Marine?" Then an appeal to potential guilty consciences! "If you are buying enough War Bonds to look him in the eye, then send him a V..._ MAIL letter. Do it today and you will sleep better tonight."

The appeals could certainly get emotional as this one from Pitney-Bowes Postage Meter Co. reveals:

> To men getting ready to fight and die, a day can be an awfully long time, a year seems like a century. Millions of men have been overseas so long they almost forget what normal living is like! So every letter means more than you can imagine – a handshake from home, a sign that somebody remembers, evidence that a good world still exists...Write often, to everybody you know overseas. And write V-Mail.

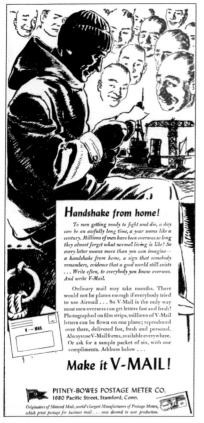

The ad concludes with a reminder that V-Mail forms are available everywhere and goes on to offer "a sample packet of six with our compliments."

Clearly, a variety of businesses were engaged in encouraging the use of V-Mail. We've seen quite a variety of art forms employed in the ads above including drawing, painting and photography. The Drackett Company – makers of Windex and Drano used stick figure drawings of a young woman being advised in each sketch "how to cheer up your soldier (sailor or marine)."

Another example of the advertising world's impact on V-Mail inducement is this most comprehensive appeal that was, "One of a series of advertisements sponsored by *Hiram Walker & Sons, Inc., distillers of Imperial Whiskey*." Virtually every aspect of V-Mail was accentuated in this ad from the drawing, complete with shadow on the ocean,

HOW ADVERTISING AND ARTWORK POPULARIZED V-MAIL 155

of the plane that could, "...make a letter hustle overseas," to the comparisons, first of a package holding 1700 V-Mail letters to a pack of cigarettes (they certainly were receiving their own endless advertisements and major usage prior to the warning labels, etc... of ensuing years, so the comparison certainly was apt) and then between plane and ship travel. It goes on to describe the V-Mail process you are now familiar with as well as all the advantages put forth by an array of companies, the government and even the Red Cross. It includes quotations by General Eisenhower and General Somervelt.

Now for our final advertising example, a short, sweet and powerful image that evokes the memory of the pigeon post seventy years earlier. These and additional images advertising V-Mail have all been located on the Duke University Digital Libraries with a search for "v-mail."

Now for some examples of V-Mail letter art. One of my favorite sources for words and images pertaining to Airgraphs and V-Mail is at this website: http://alphabetilately.org/V.html. It presents the quandary about precisely how the artwork occurred, and who deserves credit. The author of the website entry presents three pieces of V-Mail letter art and writes:

> All look to me like forms that the sender purchased and then signed and mailed, like a Hallmark card today. I am looking for further information about how that was handled - did the individuals who did the art work copy and market their own products, or was there some formal system for it?[213]

I have not found an answer to his query, but that shall not keep me from sharing some of the images the author focused attention upon in the website. The first is from Captain Ben Nathan to his family.[214]

The message, that you can wish family on the home front happy holidays and little else, is in the form of a spoof of censored letters as the implication is that the cutout parts might give away critical information.

The holiday theme permeates these next few artistic V-Mails. Not only are the greetings contained therein, but more than a glimpse is given of the location of the senders. The censor symbol in the upper left corner shows that it passed muster as drawn.

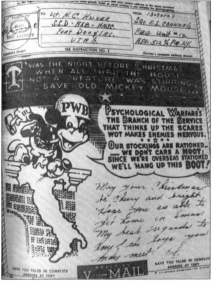

HOW ADVERTISING AND ARTWORK POPULARIZED V-MAIL 157

A final image, an anniversary greeting from W.T. Gmahle to his wife, is quite touching and definitely conveys a sense of longing in words and picture.[216]

The author of the V-Mail section of the Alphabetically website concluded his commentary on three of the above V-Mail letters with these personal comments:

> *In a way these are more poignant to me than some of the personal messages. There is more of a sense of haste and secrecy – one written six months before D-day (June 6, 1944), and with the note "Somewhere in Britain", the other written six months **after** D-Day, with the note "Somewhere in France". The last one is dated seven weeks after the surrender of Japan, when both the sender and the recipient knew it was only a matter of time before they would be reunited.*[217]

Thanks once again to my benefactor and inspiration, Bill Streeter, we have access to several images by another V-Mail artist, Edward Reep, as well as his musings, which he attached to the material he sent Bill when he was seeking V-Mail letters for a book he hoped to author.

The first letter image is best described by its artist:

> *I detested the Nazis and their psychopathic barbarian, Adolf Hitler. Hence, as I painted at the front I also wanted to destroy the enemy, and at times my artwork enabled me to join the action. My inability to successfully draw comical characters, or to "cartoon" is revealed.*

I disagree with Mr. Reep on both counts. See what you think of this drawing, entitled "If Ike Proclaims..." that depicts an artist about to clobber a German soldier with his palette![218]

Reep's next picture above, entitled "Guess Where I Am," features a pyramid and was accompanied by this anecdote:

> Perhaps the best of my V-Mail letters. I was able to let my family know where I was, presenting the message with graphic sureness and simplicity. The censors let the letter go through.[219]

We have already seen an image capturing the isolation and loneliness of a newly married husband. On the next page is another and the accompanying words convey the sense of longing and disappointment. Reep acknowledged all of these feelings in the words he wrote:

> A precise, detailed account of my loneliness. My bride and I were together for a brief two or three months before I was shipped overseas, and I (obviously), missed her so much. On December 9, 2002, we will have been together for 60 years.[220]

Finally, even though Mr. Reep didn't send his rendering of the picture that he describes in the V-Mail below, were it not for the letter he was writing at the time, he would have clearly been lost to us along with his artwork.

This V-Mail is an unfinished letter that was never posted. It was found among my memorabilia. The explanation follows.

The Anzio Beachhead movie theater, a sandbagged tent was showing "Going My Way" that night. This was a very popular movie among the soldiers, as it starred Bing Crosby. The theater was about 20 yards from my dugout or foxhole. Since I had seen the movie, I decided not to attend but to stay in my underground home and write a V-Mail letter to my family. It was the wisest decision I could have made.

As one can see, I was half-way through the letter when a Jerry 155 mm. howitzer shell landed between my fox-hole and the rear of the theater. Many soldiers attending the movie were killed and many more were wounded. It was a great tragedy. A second shell followed moments later in the identical location. It was a hot night. I had no clothes on and now I was covered with dirt and debris, with my ears ringing. I stayed there all night, frightened and a bit dazed. The following morning I painted what was left of the theater, now with men clearing the area and recovering effects of the wounded and dead. The painting I completed at the scene remains as one of the best efforts I produced during World War.[221]

Sadly this painting was not among the artworks that Mr. Reep sent to Bill Streeter prior to his sixtieth anniversary, but suffice to say he would not have been able to stay married for six decades nor supply the wonderful images and stories that have benefited this project were it not for his having been moved to "write a V-Mail to my family" when disaster struck his company.

To conclude this chapter I will draw attention to a story told in *The New York Times* on June 18, 1944. It centers around Father's Day and begins: "American troops, taking time off to greet their fathers on Father's Day, are using a special V-Mail form designed by Sgt. William Gordon of Milltown, NY." So here we do learn the name of the artist, though sadly the article only contained a description of the image he'd created. It continues:

> *Sergeant Gordon, son of Mr. and Mrs. George E. Gordon (in the days when marriage meant a woman could lose both her first and last names), drew a picture of a doughboy saluting a picture of his father arising from the smoke of a campfire as a greeting for his own parent. The drawing was brought to the attention of Lieutenant General Mark Clark, who approved its use for the entire Fifth Army. In describing the incident in a letter home, Sergeant Gordon said: "I happened to be making the design for you when it was seen by another member of the battalion. He asked if he could borrow it for a day or two. It ended up that he had taken it to Fifth Army headquarters, where it was accepted and is being printed for distribution to all Fifth Army troops in Italy.*[222]

Could this have been how the other letter artists' work was made widely available? We are left to wonder, though this poignant story honoring one soldier's father on his special day certainly adds to the mystery.

CHAPTER SEVEN

The Voices of V-Mail

One final word of appreciation is due to Bill Streeter for the contents of this chapter. Through eBay and other sources, Bill managed to procure hundreds of V-Mail letters that I have scrupulously pored through for purposes of giving voice to those millions of Americans who took advantage of V-Mail to correspond with loved ones during the terribly difficult war years. Not only did Bill obtain copies of actual letters, he also took the time to highlight quotes from those he deemed most indicative of the times. The sections he deemed worth sharing were filled with the emotions, the relationships, the struggles, the enormous challenges to which both distance and tremendous uncertainty gave rise. The hope is that providing you with the voices of V-Mail will reveal how the letters provided solace, connection, encouragement and hope for countless people on both ends of the correspondence.

I have included a number of V-Mail letters that were written to Henry Streeter, Bill's beloved cousin, in this final chapter. The letters were the ones that were returned to his family that they had sent to him, but he never received because of his passing.

There are also a few V-Mail letters that were written to the office of General Douglas MacArthur. The letter writers were a man and a woman who were desperately seeking information about missing soldiers and their last resort was to write to the Supreme Commander of Allied Forces in the Pacific.

When I was reading these letters I was transported back as effectively as any time machine we've imagined over the years could have done. Not only were the letters profoundly moving on a regular basis, but there were also exchanges of letters between lovers, husbands and wives, mothers and fathers and their sons and these brought the people to life. Reading about such letters or seeing films in which such relationships are depicted can only take one so far into the world of the real people who lived the hardships and the fears that being in a war engender. I feel it was a true privilege to enter into these correspondences and to glimpse the anguish, the love, the devotion and the boost to morale for which these V-Mail letters provide a forum.

Those who were not able to complete their thoughts and expressions of caring in one V-Mail form found ways to continue with second and third forms mailed at the same time to provide continuity, but for the most part an incredible amount of news, expressions of concern, weather reports, updates on relatives including young children and assorted views on the war and the effects on the homefront were exchanged on one V-Mail form with few complaints about the limitation.

I've organized this "Voices" chapter by year and by subject. You will see the ways in which the passage of the war years affected the types and content of the letters. You will see the variety of subjects and their expression. Some voices occur frequently and others are isolated, but the overall effect I am hoping to achieve is for the reader to have something akin to the experience I was gifted with – of what it was like to write a letter to someone you cared deeply about when there was such a thin line between life and death.

Here are the Voices of V-Mail that I selected with the assistance of three dear friends, Steve Trudel, John Berkowitz and Ken Hahn who agreed to read the excerpts I chose to help to determine which ones would have the greatest impact more than seventy years later. I have corrected grammar and spelling issues to enhance comprehension and avoid distraction.

1942

Enjoyment of serving in calm moments – Nov. 18, 1942 – Cpl. Robert Ellsworth – "I don't believe I told you that I've been issued a bicycle. I've had a lot of fun riding around the countryside – a little different then in your day, Henry (reference to World War I, most probably). I'll bet you get a big laugh out of this." (Letter sent from somewhere in England based on A.P.O. address).

Advice to parents – Nov. 21, 1942, from Cpl. Robert Ellsworth to his parents – "I'll bet everyone is rushing like mad these days. Don't forget to take it easy and let somebody else do some of the rushing, too. There is no use in getting run down just because we are fighting a little war."

A soldier feeling good about what he's experiencing – Nov. 21, 1942, Cpl. Ellsworth letter continued – "I've really seen a lot of country in the past few months. I've never regretted a bit of it. I'd probably never have gotten the chance in any other times."

V-Mail being encouraged for soldier letter-writing – Dec. 1, 1942 – Cpl. Robert Ellsworth – "Mail is starting to come in a bit faster now. Minnie, I'm able to get hold of plenty of these V-Mail forms. In fact, they urge us to use them."

Censorship, missing loved ones and expressing appreciation of wartime experiences – Dec. 12, 1942 – Commander P.S. Rudia aboard USS Dixie – "I find it very difficult to write interesting letters due to restrictions… I am having an interesting time, but being away from Lucille and the boys is a real sacrifice. Aside from that I wouldn't miss this experience for anything. Meeting and talking to the officers and men doing the fighting is really an inspiration. They are all outstanding men and are to be admired beyond words."

Difficulty when V-Mail is not legible and air raid black outs in England vs. U.S. – Dean Hill, American Red Cross, Field Director, Dec. 13, 1942 – "I have received a lot of V-Mail but never one so faint as yours. I don't suppose I can blame you for that, but must blame the V-Mail man. I can just make it out and that is all. In fact, I have to strain my eyes. But eye strain or not I'm mighty glad to get newsy letters from home… You tell the air raid wardens for me that they don't know what a blackout is until you've seen one over here (in England). The other night I got lost coming back to camp and walked eight miles instead of the usual four and over a road I am familiar with."

Censorship – Dec. 26, 1942 (Cpl. Ellsworth) – "Sorry I can't tell you where I spent Thanksgiving and Christmas, but it will all come out in the wash."

Concerns about safety of a fellow soldier – Dec. 31, 1942 (Ellsworth) "I hate to hear that Bill is going into the Army. I suppose it was inevitable, but I still hate to see it. I hope he never has to leave the States."

1943

Humor (Ellsworth, Jan. 30, 1943) – "How are the new teeth, Henry? Have you got a set for strictly formal wear, too? Ha! Ha! I got a big kick out of that one."

Work on the homefront – (Ellsworth – Jan. 25, 1943) – "So you and Mom are both thinking of going into a defense factory? Labor shortage must be getting pretty tough."

Missing soldier – letter from H.C. Hicks, American Red Cross Local Chairman in Gilman, IL to Gen. Douglas MacArthur – Feb. 15, 1943 – re: Pvt. Charles Edward Ohler

Dear General:
We have tried through every Red Cross Channel to locate a boy for a distracted mother, a boy who was most likely in the Philippines on Dec. 7, 1941 (Pearl Harbor attack day!). All gov't. inquiries report "no trace of this Company." He wrote regularly and sailed from Hawaii on Pres. Coolidge and when he wrote last on Dec. 1 1941 he had landed in Philippines and was on a Field there and expected to be moved soon, but did not know where. All mail and Xmas packages and Red Cross

kit sent him since that date have been returned marked, 'No trace of this Co.' Can an entire Co. be lost with no trace of whereabouts? He is the eldest son of a widowed mother – & another son in service and this uncertainty is very hard on her. Is it likely he is interned – or if a casualty would War Dep't. have no way of knowing?

The people of the U.S.A. have great confidence in you – and most of them pray each day that you will soon be given all the equipment you need to accomplish all that we know you deeply desire. If you can tell us anything to tell this mother we will deeply appreciate it…

Missing one's family – January 25, 1943 – Lt. Commander John Oakley, Marine Corps Unit 845 – "I hope you and the children are fine. I think of you constantly. Your pictures have been such a consolation to me. When I get more permanently located I want some more pictures. Tell me all the little things they do as I am anxious to know all of their progress. I hope my big girl is still doing well in school. Her daddy is so proud of her. I love you dearly darling and am living for the day of our reunion. Many kisses and love to the babies and you. Bud."

Feeling distant, expressing a sense of isolation, acknowledging how much worse it could be – March 19, 1943 – "Birthdays and that sort of thing cause me to reminisce a bit. It seems so long ago to my twentieth birthday that I can hardly remember it. However, I have no need to kick, for across the barracks from me sits a fellow who is writing a letter home to his family. He received word today that his mother had just died."

Censorship – May 24, 1943 – Sgt. Ellsworth – promoted to sergeant – "It's xxxxxxxxxx the very xxxxxx xxxxxxx. It looks like it will keep up all day. We had been having some nice weather, but I knew it was too good to last."

Rationing – April 1, 1943 – Sgt. Ellsworth – "I saw a newspaper article which said they were going to stop the sale of gas in the eastern States altogether, except of course of absolute necessity. You people must be having it pretty tough right now and if they do that you will have to walk, without any question. You had better buy a horse, Henry. You may be suffering from rationing and restrictions, but I hardly imagine that you have experienced anything like the British have for five years. So you will have to 'bite off another Chaw' and 'dig in'. Regardless I hope this doesn't have to go for too many more years."

HISTORICAL REFERENCE – *"Doolittle Raid on Tokyo"* – Roosevelt was hell bent on attacking Japan following Pearl Harbor so sixteen planes dropped bombs on April 18, 1942 – five months after the attack on Pearl Harbor. One mistakenly fell on a school causing a great uproar throughout Japan. Eight crew members of the downed planes were captured on mainland China and three were "accused of indiscriminately killing civilians." All were tried for war crimes and sentenced to death. The Japanese executed three in Shanghai in October 1942, but commuted the others' sentences to life in prison; in part for fear that executing all of them might jeopardize Japanese residents in the United States (later interred in camps anyway). Japan retaliated with major attacks on the coast of China where American planes sought refuge following the bombing.[224]

The Doolittle Raid was written about in the Ellsworth letter dated April 24, 1943. "I see by the papers that they executed those men who took part in the raid on Tokyo. That was a pretty raw deal wasn't it? I don't blame the American people at all for getting all worked up over it. I personally would like to see them get something in payment for that. In fact, I imagine everybody would."

Loss of a finger and imagining the impact when he returns home – April 27, 1943 – Corporal Art Hammer – "Well, I am now in England once more and it was quite a rough job getting here. By the way, find out from T.O. Moloney if he can use a man with only nine fingers when this war is over. I hope so or else I may go and find myself a corner and go to selling apples or something. But outside the loss of a finger I am in good shape after more than four months in the hospital and the prospects of a few months more."

Hardships and homage to the coping ability of a spouse – May 31, 1943 – Cpl. Carlo Lozano – "I am proud of the admirable serenity in which you've taken possession of the events of my current circumstances. With positive, unimpeachable gaiety; I rejoice that you are so brave, dearest. Doubtless it is the very note of praise, of lavish praise that I must evoke for your reactions and undying loyalty. May God bless you!

Here in N. Africa the heat is intense. – from the same Lozano letter – "Things that we take for granted in our home, such as an ice-cold drink of water, is a luxury, or better, impossibility. A 'coke' is a thing out of the other world. But we have to bear with high morale and grim determination the tragedies of current events – so that in the future we may be permitted without interference, to resume our normal luxuries so many other people of this world are denied."

How a soldier died whose body is missing – a letter to General MacArthur from Mrs. Martha Hickok, sister of Lieutenant Frank Thompson, from Miami, FL, June 13, 1943

Dear Sir:

Could you please give us some details of the passing of my brother, Lt. Frank Thompson, 38th Bomb Group. We have received the official notice from the War Dep't – also the fact that his body was not recovered. We have written to his friends, his commanding officer and chaplain for details and have received no answer from any of them. And yet I know a woman here who has received four letters from her son's C.O. (Commanding Officer) telling her how his plane crashed and just how her boy died. I know others who have received letters from friends who were with their boys on missions. All of this makes us feel as tho' no one actually knows and makes us wonder if it could be a mistake. Can you tell us since his body was not found? What proof is there of his death? You can understand our frame of mind. We cannot start to forget until we are sure. Will you please answer this and tell us some details so we can be sure?

Lt. Col. Morehouse, MacArthur's Aide-de-Camp, responded to the above V-Mail and then Mrs. Hickok wrote again on July 21, 1943.

Dear Colonel Morehouse:

I just received your letter in regards to my brother, Lt. Frank Thomson. I wish to thank you for your prompt reply and only hope that this reaches you. I think the reason I am still trying to find out the details of my brother's death after eight months is that not all of them are received.

You made me feel very optimistic for a minute when you said you knew that General MacArthur joined you in hoping that my brother is alive and well. Then I thought you hadn't actually seen my letter telling you that we had received word from the War Department the 17th of February telling us that he had been killed in action Dec. 5th. We received word from the Adjutant General telling us that he was on a bombing mission when the plane crashed and the entire crew was lost; however, his body was not recovered and security regulations prevented telling us any more as to where or how the accident or crash occurred. What we want so badly to know is: what evidence was found of his death since his body was not recovered? Were there remnants of his body, but not enough to recover and if not how do they know he is dead? We were told that part of their identification tag was always brought in to their base and then sent to Washington with their belongings as evidence of their death. Can you tell me if this was done with my brother's case?

I believe you can understand how we feel when there is a doubt in our minds. We haven't even received his belongings and there are other things that give us some hope. Of course, they are probably trivial and we magnify them, but on the other hand the fact remains that after 8 months of correspondence I still don't know what happened to my brother.

If I wanted to know something that would violate security regulations it would be a different thing – but I only want to know what evidence was found of my brother's death. I will be most anxiously awaiting your reply and I do want to thank you for what you have done.

The final attempt by Mrs. Hickok to find proof of her brother's fate in a letter to Lt. Col. Morehouse following another letter from him.

Dear Col. Morehouse:

I only hope this reaches you as I have misplaced your address. This is in answer to your letter, which came through the War Department. I wish to thank you for the information you were able to give us and ask, if I may, one more question: would it be possible for you to communicate with my brother's base or in some way get tough with those who found the wreckage and find out if his body was positively accounted for as since you told us that his plane crashed with a full bomb-load it seemed to us that there might be a possibility that the bodies would not be there to be accounted for. If that would be the case, then it would seem possible that even if one of the crew did escape before the crash no one would have any way of knowing about it. Which all comes to the point of that which I haven't written before for fear it would be censored, but on careful thought I fail to see where it would violate any security regulations so I will tell you one of the reasons we are asking for further surety of my brother's death: a middle-aged woman living near the field from which my brother went overseas wrote to my mother in Calif. telling her if she could see her she would tell her something, which for some unknown reason, she did not want to write. My mother then traveled across the States and found it to be that this woman had overheard high-ranking Army officials say that my brother was not dead, but a Japanese prisoner. This was to have occurred at a dinner party. She, this woman, was asked not to repeat it but since it happened months ago I feel that only good, instead of harm can come of telling it. At first we were inclined to disregard it entirely and did so, but for one thing which she repeated which she could not possibly have made up and which, to an uninformed person, would sound silly and which we later learned was not silly, but entirely true. This, then, made us write many letters to find out the minute details of the crash. We believe that you are the only one who can answer such a seemingly trivial question as to exactly what evidence was found of his death, since you would have to be in a position to communicate with those at the base.

I hope this is all clear to you as I have condensed it as much a possible so as to get it all in this letter. If I do not hear from you in a reasonable time I will duplicate this as I know its address is inadequate.

P.S. Another thing that gave us hope. After 8 months we have not received any of his belongings.

Seeing the world of opportunity afforded by military service – July 31, 1943 – Cpl. Carlo Lozano – "I left Africa – Africa, the land of mystery that spreads from the beautiful, blue-green soft waters of the Mediterranean to the dark depths of jungle moss. You should see the multitude of races – from French to Hebrew, Arabic and Spanish. Acquaint yourself with the Northern map of Africa, especially the bigger towns as Oran, Bigerte, Tunis and even Carthage. I have heard fascinating tales of such places, which I would like to relate to you when I come back."

Fighting [and getting past the censor who signed off maybe due to the letter writer's rank!] – August 8, 1943 – Major Vernon Watkins from Tunisia, N. Africa – "We had some pretty tough fighting during the campaign. In all we were in constant contact with the enemy for 3 ½ months, with the exception of that move to the north. We were out for a week. Our losses were moderately high, but I got thru without a scratch though had plenty of close calls to suit me.

Once more I thank God that you are not in want or hungry. I have seen a lot – of what the tragedies of war have brought upon humanity. Now more than ever I gladly give my life to spare you of the pathetic and pitiful and everyday occurrences of this frightful existence. May God permit that this war shall never be brought to our land."

Growing up fast in war – September 6, 1943 – Sgt. Robert Ellsworth – "In a way it doesn't seem as if I have been away from home for so long. In other ways, it does. I sure have a different outlook on life - you can bet on that. Possibly it is because I am a year older, but I am more inclined to think it is my associations and experiences of the past year. The generation of today has progressed ten years in two years of war."

London is full of "Yanks"; wishing for winter on the Riviera… – September 10, 1943 – Dean Hill, Field Director, American Red Cross – "I haven't been down to London for fun in months. I've been down a few times, but came right back on the same day. (Business) It is terribly difficult to get a hotel room or find a place to eat in London these days. The place is overrun with these bloody Yanks.

I bet you have your hands full with two jobs and I bet you get mighty tired. That's why I appreciate it all the more when I get a long newsy letter from you, as I know you're darn tired and would rather go to bed than write to me… I am well, but am looking forward to a lot of colds this winter. I wish they'd send me to Italy for the winter. I always wanted to winter on the Riviera."

V-Mail letter speed, local culture in Sicily – women's shoes, language, children – September 11. 1943 – Cpl. Carlos Lozano – "Your regular air-mail letters took over a month. But others have taken less – minimum 23 days. V-Mail letters from 12-15 days if your dates are accurate… The shoes you gals are wearing now are similar to what the Sicilian girls are wearing here, only their sandals have wooden soles and you can hear them coming for blocks. You ought to hear them talk – fast and furious. I can't even count the no. of words per breath. But they always ask for something of us, especially sweets – or as they say "caramela"… You ought to see the bambinos. They are learning English fast. We are approached sometimes with, "What's cooking, Joe?" And then of course the never failing "caramela". They have learned our G.I. terms. Oh, but Arab kids are the smartest. Some of them talk Italian and, of course, French and their own Arabic language. How they picked up our army talk was marvelous, but it isn't spoken in the best of circles. Darling, this country is full of Homer's Greek mythology. Of the Roman conquest. Some of the cathedrals are masterpieces and the architectural beauty dates back to the 13th century. 'A thing of beauty is a joy forever' as Keats once said."

Speed of V-Mail – November 14, 1943 – Pfc. Bruce Cole – "I got an honest to goodness letter from you yesterday written when Nan was there with you. The V-Mail written the same day got here much sooner than the regular. Dad, I bet you are still knocking the deer over."

1944

Importance of receiving mail for a soldier – January 15, 1944 - Rudy Kolenic – tells how to address a letter and then writes: "All of that has to be put on the letter or my mail will be delayed for quite some time. So please put the correct address on my mail as I hate to get my mail late. Mail is one of the most important factors in a soldier's life. A letter can do anything to a man."

Packing a lot into a V-Mail and the Russians impact on the war – Sgt. Robert Ellsworth – January 10, 1944 – "Received your V-Mail letter Dec. the 26th, today. You sure can pack a V-Mail full of news, Minnie. I'd say you were the champ in that respect…It sure looks like the Germans are taking a beating now. Those Russians sure are a tough bunch. They deserve a lot of praise and consideration."

Pride of a loved one on the home front – March 20, 1944 - Marion Craig – "You are now receiving letters from the wife of a Marine. There were 1,600 men at Leavenworth and MY Chubbin was one of the 27 men picked for the Marines. You can bet I am a little bit proud. They are attaching officer's candidate papers to his regular induction papers. And, as badly as I hate to give him up to the Army as anything else, there is a sense of pride inside me, which would be equaled by no other feeling in the world."

Feeling disconnected from home – March 8, 1944 – to Mildred Prince from a Navy man – "Can't tell you where we've been, going or what we're to do, but if things keep going the way they have been…. It won't be long now. We seem to be lost to the outside world – no news or anything. We don't know what's going on back home and when you've got a few minutes… you start to wonder…."

Commentary on quality of draftees/recruits by April 1944 – to Pvt. Roger Bettig. "When I came in we had to be a man in all respects to get in, but now they just look in your ears and if they don't see any light they feel you and if you are even slightly warm they put you in the Army."

Not receiving mail on the home front – March 14, 1944 - W.D. Robertson (Dad and Mom) to son at war – "It is raining here today and the weatherman ways it will turn to snow tonight. I just got back from town. I bought a tube for your radio. It is O.K. now. Well, son, we haven't got any letters from you since the 25th of January but I heard on the radio Sunday night there had been 800,000 letters come in to California from the Pacific area. Out of that many letters there ought to be one from you. I sure hope so anyway… This is the first V-Mail letter I ever wrote… Well, son, I will close for this time. You write as often as you can. Love, Mom and Dad.

Hatfield, Massachusetts farm family – April 15, 1944 – Cpl. Edward Kochan April 15, 1944 – "Well I suppose by now you and dad must be well underway on the farm. What are you raising this year? Any more of the guys from town going in the army or are they all in already? We have been having that good old Hatfield setting onion weather here the past couple of weeks, as it's really been nice out."

Discipline, "misdemeanors" and fines – April 20, 1944 – Pfc. George Gross – (written to Mrs. Gladys Gross of 124 Cottage St., Easthampton, MA) – …"This time it's a honey. Guess they don't even want us to go into town. There are M.P.'s moved into all the towns around here now and on order of some big shot they'll pick you up for the least little thing. The drive started yesterday and the very first day eighty fellows were picked up. Some of the misdemeanors are: button open, one or both hands in pockets, hat worn incorrectly, unshined shoes, uniform wrinkled, etc… And the best of all, the hair can't be longer than two inches. Now for the fines: button open – $5, one hand in pocket – $15, both hands in pockets – $25 etc… And for each soldier that is fined the company commander of his outfit has to pay the same amount – he definitely won't like that."

Morale boost of letters – April 26, 1944 – Cpl. Carlos Lozano – "Always know – you must know – what your letters mean to me. They are 'LIFE' – understand, 'LIFE'. I only live in hopes – waiting – waiting for your sweet words. Don't ever fail me. I know you won't. You have always been so wonderful – be thoughtful – so sweet and so sympathetic."

Missing home – May 15, 1944 – Pfc. Bruce Cole – "We are listening to the radio tonight and there is some swell music on. Kind of makes me feel homesick, but I have been away from home so long now that I am getting used to it. They are now playing Rudy Vallee's theme song and it certainly brings my memory back to good old Maine, which I hope I will see some time in the near future."

Censorship – May 15, 1944 – Capt. Leonard Lord – "I cannot tell you anything about my work although I do not see why."

Saving space in a V-Mail letter – May 16, 1944 – Cpl. Geroc J.A.S. to Miss Kathryn Streeter, W. Cummington, Massachusetts – "I'm not using paragraphs in this letter, because I want to save space."

From the Asian theater to the European theater – May 11, 1944 – Warren Lyman Barry, naval officer to Capt. Lyman Barry – "At present I have around $50 riding the books and I let the majority of my pay always ride, since a month or so ago. I will try to get you a Jap souvenir, soon as I get my first Jap. I always wanted a small German Luger, so is there a special Jap token that you wish? Enemy currency is quite a token for some. I intend to leave much of my stuff behind if I have any action…"

News of a mother's death – May 13, 1944 – Pfc. George Gross – "That Tourville fellow finally got the news today that his mother died. He was in front of the guardhouse ready to go on patrol when his lieutenant came over and told him. He took it pretty good, although he did cry a little. I'm glad I didn't tell him when I got your letter. P.S. He was taken off guard duty for the day…"

Hope for the best – July 26, 1944 – Sgt. Robert Ellsworth – "I haven't seen Jimmy Dorschel or heard from him since I saw him last. I've been kind of wondering what's happened to him. After you have been over here for awhile all you can do is just hope for the best. You learn to expect a lot of things. I'm sure he is all right though. Probably just too lazy to write."

Censorship – July 21, 1944 – Sgt. Keith Rodney, Jr. – "After writing home several times recently of my whereabouts, I now cannot do so. I suggest that you get in touch with the family. Until further notice my whereabouts is not only a secret to you and me, but to anyone as far as I can see."

Frustration with present location – July 27, 1944 – Sgt. Warren Barry –"Life here would be very fascinating to watch in the movies, but actually living here is hot and allows only the professor of human nature, to see the comics of tropical G.D. (God damned) life. (I hope you still have your magnifying glass? – (to read the V-Mail letter.)"

Morale boost of mail – August 29, 1944 – Pvt. Raymond Ellerson – "Let's hope and pray for it all to end soon. Mother, I will bear it for your sake and I don't want you worrying about me. After all you know I am pretty rugged and able to take care of myself. All I want is your mail and news of you and daddy and home. This will help more than you know."

Boredom – August 17, 1944 – Cpl. Carlos Lozano – "The monotony of this place is slowly affecting my nerves. It is too quiet. No air raids, no bombs bursting close by. No sounds or screaming shells."

On the homefront – August 15, 1944 - Mrs. W.P. Shortell – (2 months after D-Day I invasion). "I got up at 6:00 a.m. today – couldn't sleep – turned on the radio and got news of the 4th invasion. I just hope it gets over and quick with few casualties."

Reflections on serving in France/impressions of the countryside/scars of war – September 24, 1944 – Pvt. Rockwell Gardner– "'La Belle France' and so she is tho' the nights begin to get pretty chilly. The countryside is very green, houses are almost without exception stone and the scars of war abound (as you see by the papers). Of course, the damage is spotty. Many houses are virtually untouched while others are in ruins… Have talked just a little with the natives. Get along fairly well. Peaceful enough in the apple orchard where we're encamped."

Yom Kippur greetings and expressions of intimacy – September 27, 1944 – Bil Shortell, Jr.'s wife, "Butch" – "And a happy Yom Kippur to you, too… Mmmm – three V-Mails from Pvt. Bil - !!! I'll bet you could hear me yip all the way over there – and Belgium it is now. And thanks for the tip on how they were speaking Flemish where you were – now I can pin you down in one spot more or less – of course, you'll keep rolling along, but by following the papers I'll catch up with you. Honey, I especially enjoyed the letter you wrote September 17 – 'a quiet Sunday in camp with church bells ringing in the distance – and at that moment parachutes were drifting down into Holland… 'You know what I've been doing since the mail came? I've been staring at your picture and trying to put a mustache on it. I'll be, it looks real purty. Will it tickle, honey? ('Scuse me a moment while I scratch. The power of suggestion has me twitching). Before you change your mind and hack it off again what is it like? Like so? Or so? Or so? Leave me know, my love. I pant in anticipation. I LOVE YOU. Forever yours, Butch."

Report from the homefront about German surrender and single-handed capture of 56 Germans! – September 4, 1944 – Mrs. W. Shortell to Bil Shortel (husband of Butch!) – "A special news came over radio this a.m. that Germs had capitulated to the Americans – not official picture in last nites paper of Olin Dows, the man who captured 56 Germans single-handed. He is from Rhinebeck, N.Y... Good old Duchess Co. blood – artist by trade. I'll save the paper for you – also some of the big headliners of recent vintages. In case you haven't heard the story, Dows thought he was rounding up a handful of Germs and to his surprise 56 popped out with hands up. He convinced them in perfect German. I hope very soon you will be permitted to tell us where you are. More news as I get it."

Missing/longing for one's wife – Oct. 29, 1944 – Cpl. Carlos Lozano – "You do look too glamorous. Could it be that you are all mine! Take care of that 'chassis' for I am going to take you for a ride – I mean on a jeep. How about a snap-shot in a bathing suit or shorts or… I want it for my pin-up girl. I can look at you before going to bed & have more beautiful dreams. Yes, my dear, I received the pictures of all the kids."

Encounter with French Cure/priest – Oct. 2, 1944 – Pvt. Rockwell Gardner – "Talked with CURE last night. A good man. Prisoner-of-war in Germany, 1940 'til recently. Doesn't speak English. His church has roof blown off all except tower. Heinies (a derogatory term used for German soldiers that originated in World War I) thought American parachutists were hiding in it, so boom-boom!! Parts of it very early…Haven't we a small French-English and vice versa dictionary around? Please send it if so…"

Horrible battle for Palau Island – October 1944 – letter from Pvt. J.M. Hayden to Kiwanis Club, Middleboro, MA – "This is our third blitz and I hope to the Lord it's the last. Twenty months at the front now... They say we have priority 1-A on going back to the states after this job. I have a galley all set up here now. I am on Palau Island. You will be reading this battle in the papers soon. It was the toughest yet. Japs all over the place. They had some of their imperial crack troops here. And we lost quite a few men that were dead in the water and all over the place... The Japs had a mt. on the Island. All stone with tunnels running all through it and it was hell to get them out. They were dug in like rats, pill boxes, 3 ft. cement walls, 35 ft. deep, doors 14 inches thick so you can see what a tough job it was. Say hello to all brothers. Best regards."

Conditions in France, malnourished children – October 4, 1944 – Pvt. Rockwell Gardiner – "Anyway 'tis a nice countryside and without the war clouds people could be happy enough in their stone houses. The children seem fairly happy, but some are undernourished. I'm pup-tenting with a chap from B'klyn. We were in a home last night showing family photos of our nieces. DeNie has one, too. There were two small boys and a girl about 10. She was studying her schoolbooks. Nice children. They were like our children. They'd been eating cocques (cockleshell mussels), which are brought from the sea. Food in a varied diet is not too plentiful for them. We swapped them an orange and le pere said one of the little boys had never tasted one."

DeGaulle is mentioned vis a vis his Notre Dame visit [into Free France following successful D-Day invasion] – October 5, 1944 – Marion Shortell – "I have been listening to the radio quite a bit lately and I got a delayed broadcast from Paris last nite. The commentator spoke about DeGaule (sp.) and his trip to the Cathedral of Notre Dame. It was interesting. I hope that if you get near Paris or other famous places that you will visit some of the old cathedrals so that you can tell me about them when you come home."

Family member (father's) condition on home front – October 1944 – Mrs. W. Shortell – "If you worry about Pa, you don't need to – he looks fine and don't do a top of work – only fish and carry home groceries. He takes beautiful long trips thru the country, had a good time – free and he is well-tanned. Our bathroom is now a "keeper of bait" line things and awful to look at. I guess they are polly squirmers or wigglesticks or somupi. Anyway I use a pail now, taking no chance of one of them things jumping up at me."

A soldier at the Front seeking news about the war…from home - November 4, 1944 – Pvt. Raymond Elberson – "When you write please tell me all the news about the war. I don't hear much. You know more about it than I do".

Foxhole life and driving a G.I. truck through France and Belgium – November 10, 1944 – letter from PFC Roger Hills to Pvt. Henry Streeter (to whom the book is dedicated…) – "I'm on the Front now. Shells are going over most of the time. We are living in foxholes with dirt and logs over us. They are all-right until it rains and then everything gets wet. It is still fairly warm over here. Once in a while it gets a little cool… I drove a G.I. two and a half ton truck over here for a whole hour. I was able to see most of France and part of Belgium. We traveled as fast as the trucks would go – usually about forty to fifty miles – until someone put up a fuss and then we had to slow down to thirty and thirty five."

Missing his wife – November 23, 1944 – Cpl. Carlos Lozano – "All day long I have had a terrific case of 'blues'. From your letters you have an attitude, or should I say, formed an impression that we are having a wonderful time here in Italy. My darling, I am sorry – but you would be greatly disappointed. It is true we have a dance about once every 20 or 30 or so days; we have USO shows; we have pictures of more recent release. BUT – how can anyone enjoy anything when his spirit is low – his heart is broken. I don't seem to find joy in anything I do. My heart is not in it. Have you ever gone to a dance and derived no enjoyment because your heart wasn't in it? You could not possibly get any enthusiasm because something was missing. Have you ever danced, mechanically, yet your mind and your heart was somewhere else? Have you? Have you ever gone to a show because you were lonesome? Well, honey, I have and I must admit, I AM LONESOME! They can have all of Italy and France thrown in for good measure and I would only be back here with you."

Christmas, missing his wife, sympathy for "fellows up at the Front" – December 24, 1944 – Cpl. Carlos Lozano – "Here it is – Christmas Eve – I am wishing I were with you. We have now an E.M. (Enlisted Men's) Club in our own building. We have liquor (Italian), which definitely I do not care for. We are to have some USO entertainers coming to our club to cheer us up. So my dear, you may imagine what Christmas Eve I shall have. But darling, who am I to complain? The fellows up at the Front are exposed to shells, rain, snow & wind. Would like to have Kathleen Winsor's book *Forever Amber* and some after shaving cream." (Published in 1944, three million copies of *Forever Amber* were sold and the book went on to become a bestseller in 16 countries. Kathleen Winsor's story of an English adventuress who becomes one of the mistresses of Charles II had been banned in Boston as "obscene and offensive". In banning the book, the Massachusetts attorney general had listed 70 references to sexual intercourse, 39 illegitimate pregnancies, seven abortions, 10 descriptions of women undressing in front of men, and 49 'miscellaneous objectionable passages. [from a review by Elaine Showalter in the British newspaper, *The Guardian*])

Marriage break-up of a soldier friend – December 20, 1944 – Cpl. Carlos Lozano – "I shall try to give you the story of Sgt. Myers. Harry and I started from Camp Robinson together and we have, from Stoneman to Africa, to Sicily & to Italy been together ever since. Harry is very much in love with his wife – a nurse that lives in L.A. Harry in all that time that we have been overseas, has been faithful to her EVEN IN THOUGHTS! He has actually dreamed, day & night, of coming back to her. He has lived only for her & to come back to her. BUT – the never failing, *but…* Some time past he got word from her or someone in L.A. Harry is very reserved & he trusts me, but all he did tell was that he was having trouble with his wife. So great is his love for her that he couldn't carry the torch. He just blew up. He went to pieces. But God is kind – Harry is on his way to Los Angeles. God grant them reconciliation for he does love her. I know. I see him, night after night smoking one cigarette after another – ALL NIGHT LONG. Honey, this is confidential between you and me… When anybody like Harry loves his wife so very much, he doesn't deserve such treatment. But who am I to judge other people's actions?"

Christmas on a military medical ward in England – Dec. 27, 1944 – Pvt. William Shortell – "Our Christmas was quite nice. The ward is garlanded with colored papers, with red and blue bells stuck here and there. We have a small tree, decorated with cotton for snow and hung with medicinal vials containing colored water. Christmas Eve we were entertained by a chorus of GI's and officers, both men and women, singing the Christmas carols. They visited all the wards. They gave out boxes containing candy, two socks and cigs."

New Year's Resolution, illness (jaundice) and a new address – December 31, 1944 – Warren Lyman Barry, soldier (rank unclear from letterhead) to his dad – "So our New Year's Resolutions are clear – to win and go home, right? Then comes the Post War… When your awaited next air-mail letter arrives, I'll answer you via air mail right away, but meanwhile I'm conserving stamps (V-Mail was supplied and mailed free of charge). Please note the new address as I'm transferred from the Sea Bees till they cure me. This is my third month of this cursed yellow jaundice and its just sapping me, keeping my eyes and skull as yellow as a Jap, yet I'm still strong."

1945

French chateau damaged by bombing – In France February 1, 1945 – Pvt. Rockwell Gardiner – "And now we're living in a town chateau built not later than 1750 and probably plenty earlier. The town is full of ancient buildings and a castle, part of which is from Roman times, overlooks it. The lived in part of the chateau has had its roof considerably damaged by artillery fire. It's now uninhabited but locked up and labeled booby-trapped."

Losing a comrade-in-arms – via returned V-Mails! – France, Feb. 14, 1945 – Lester Hawkins, parachute infantry – "Got a letter from Juanita and Noel and two from you and one from Maurine, asking about J.J. I have two letters I had written him that came back, marked deceased, so I'll send them to her. She thought maybe he was taken prisoner."

March 27, 1945 – Lester Hawkins – "Sure glad you received the shoes. They're real clodhoppers. A Dutchman in Eindhoven, Holland gave them to me while we were there… Got another letter from J.J. marked deceased so he's surely killed."

The hell of war in the Pacific – April 12, 1945 – Lt. W.A. Zerousky – "Just a few lines to let you know that things are shaping up pretty good. My forward party was under two days and nights constant shelling by the Japs, but we held the front lines. It sure has been tough as all hell. I got hit in the hand by shrapnel, but it's not bad. Shrapnel was falling like rain all around us. I was in a foxhole and I'm not ashamed to say that I prayed all the time. I still have a terrific headache. We've been up the front lines from the first day we landed. If I don't get relieved soon, I'll go nutty."

Roosevelt's death – April 13, 1945 – Lester Hawkins – "Sure a bad time losing the president – everything in the stage it's in… Saw in the paper where they're sending all physically fit men overseas who haven't been over already."

From Henry Streeter's stepmother: (letters Henry never received and were sent back after his death was confirmed – Henry is Bill Streeter's double cousin to whom the book is dedicated and whose life and death inspired the book.)

> ***Sending a Bible and warning about French women and girls*** *– April 4, 1945 –* "Just a line to tell you that we sent you a package today. I sent the Bible, which is just the New Testament, but I liked the print as I thought it was easy to read. I think you will like to read Matthew… We sent you some candles, but hope this war in Europe will be over before you get this package so you won't need to be in fox-holes. Even after it is over I think the civilians are going to be very dangerous, especially the women and girls. Be careful about trusting any of them. I just don't think you can… Jan is waiting for me to finish this so she can go to the Post Office so will sign off for now. Don't forget to write when you can. We do worry when we don't hear. Be careful. Hope you are well."
>
> ***No letters for four weeks and worry about dangerous civilians in France*** *– April 3, 1945 –* "I am wondering if you are getting any of my letters. I hardly miss a day writing to you so you ought to get some mail. Grandma is writing about 3 times a week. Gramp goes up to the Post Office about every day now hoping to hear from you. It has been 4 weeks now so it seems as though we ought to hear soon… I hope you are well and O.K. Take care of yourself as well as you can, be careful and don't trust any of the civilians over there. Arthur Streeter says the civilians can't be trusted where he has been in France. He didn't dare go out after dark. He had a bullet bounce off his tin hat and he thought that was close enough."
>
> ***Henry's stripes/medals – his modesty and step-mom's pride, money matters*** *– April 10, 1945 –* "Congratulations on your first stripe. You are pretty modest not to mention it, aren't you? When did you get that and how come? I hear it is hard to get

them over there. I think it is swell and it means more pay, too. Have you got your Purple Heart yet or your ribbon? How many ribbons have you? Do you get a ribbon for each country you go through? For example you landed in Scotland, then England, Belgium, Germany. Have you got 4? Roger wrote he had his ribbon but not the medal. While I am asking questions, remember I am waiting to hear about the arrangements you made about your money. I haven't received any bonds since the ones for the month of December. You remember when you called from Meade you said you were having two bonds a month. I don't care what you decided to do, but I am curious to know if you made other plans for I haven't received any and if they should be coming I think somehow we ought to investigate… Walt Murray is still in the hospital, but I don't know where. He was hit in the leg with a shell or part of one. What were you hit with? Can't you tell us about it? Don't tell Gramma until she is a little stronger. She is gaining. She says she is going to live until you come home. She is very persistent so I hope it is true, and I hope it won't be too long before you start back…"

Roosevelt's death, fears and anxiety about Henry – April 13, 1945 – "When we got home we heard the news about Roosevelt. We listened to the radio instead of getting supper. When Gramp was calling before the election for me to vote I told him he would never live through another 4 years, but I didn't think he would die quite so soon. I wonder what kind of a man we have for President now and I think a lot of so-called democrats are wondering the same thing. Well, it can't be much worse and I hope it will be better… I am wondering where you are. When I read the papers I get the jitters, wondering just where you are in the strife. Very little war news has been on the radio this morning. It's been mostly about Roosevelt either in memorial services or news commentators. Write as often as you can. Our thoughts are with you all the time."

5. Expressing hope to hear from Henry and passing on a compliment he received about his truck driving – April 14, 1945 – "It's been another week since we have heard from you so we are hoping to get a letter today. We keep wondering all the time where you are. (Of course, the censors prevented him from updating them with his location)… Bartlett, your old friend down at Hoods, was here yesterday checking up on warm milk. Louis Sears water was 70 when he was up there. You see most of the farmers bought coolers because no ice was cut. Some coolers have come and some haven't. Bartlett asked for you and thought you had moved along pretty fast. He gave you quite a compliment. He said you were the best driver in the milk truck he had. Wasn't that quite a lot for him to say? I have had so many interruptions that as usual my time is up. Dad has come in to dinner. So will close for this time. Write as much as you can…"

Forest fire on Mt. Tom, Holyoke, MA – letter written the day of Henry's death – April 17, 1945 – "We are trying to have a rainy morning. We need the rain, as everything is getting very dry. Over the weekend there was a forest fire on Mt. Tom. The State Guard was called out Sat. night to help fight it. Charles Cummings said the whole top of the mountain was burned over before it was put out. The fellows got back here around 4 p.m. Sunday… The news (from Europe and the war) sounds good today. Hope you are O.K. Take as good care of yourself as you can… Send me another request so I can keep the packages coming. Lots of Love, Mom and Dad"

Russians and Americans uniting against German army in Europe – Lester Hawkins – April 28, 1945 – "The Americans and Russians finally made contact – was sure glad of that. Maybe it won't be too long until they clear the bastards out of southern Germany."

End of war in Europe and what's next in the Pacific – May 11, 1945 - Lester Hawkins – "…Well, I guess everything will turn to the Pacific now to finish things over there… Guess everybody rejoiced over the end of the war with Germany. Hope the Russians go against Japan now." (Some historians believe had we given the Soviet Union time and opportunity to invade Japan the decision to drop the atomic bombs on Hiroshima and Nagasaki would not have been made!)

Non-fraternization with local people in Austria – May 16, 1945 – Kossen, Austria – Lester Hawkins – "We are not allowed to fraternize with the people so you ignore them when you see them, unless they're from some allied country, which you can tell by their identification."

"Disposed persons," food shortages in U.S., proximity to Hitler's hideout in the Alps… Berchtesgaden, Germany – May 31, 1945 – Lester Hawkins – "There sure are lots of Disposed persons around here (he meant Displaced Persons). Mostly French, Polish, Russian and some Dutch… I saw in the paper where there was going to be quite a food shortage in the States this year… I can see old Adolph's hide-out (crow's nest) from the window of my room. We are in German barracks formerly occupied by SS troops, they're really nice, too."

Status of forces update from General Taylor – Berchtesgaden, Germany – June 25, 1945 – Lester Hawkins – "General Taylor, Commander of 101st made a speech today and said we would remain here until the first of the year. Then go to the States and get at least 30 day furloughs, then be in reserve for the Pacific if they need us. Said that's the way the W.D. (War Department) had us slated now. Therefore I guess it will be about six mos. before I get in unless orders are changed between now and then."

Update on where he will be stationed – Berchtesgaden, Germany – July 1, 1945 – Lester Hawkins – "Where did Noel get the word we would be in by the 1st of June. Too bad he wasn't right. There is more than one division over here. There's the 101st, 82nd, 17th, and 13th divisions and the 82nd is occupying Berlin as the Army of Occupation and the 13th went directly to the Pacific and the 101st and 17th are going to the States to get furloughs and be in reserve ready for MacArthur in case he needs us. That's the way it stands now anyway…"

Reference to friend being drafted and to atomic bombing of Japan – Bar-le-Duc, France – August 8, 1945 – Lester Hawkins – (two days after Hiroshima bombing) "…Did Bud ever make it in? How did he like Salzburg anyway? It shouldn't be long to finish the Japs with the new bomb they're using now, should it?"

And it wasn't. While we will never know for certain whether World War II could have ended without the incomprehensible destruction of two major population centers within three days of one another in August of 1945, we do know, from these excerpts and countless others that could have been used in addition to the ones I selected, that such letters contributed to sustaining the war effort to its completion. The voices I have sought to recapture here enable us to gain a glimpse into what it was like to experience life under extremely arduous conditions for the people at both ends of the letter exchange. It is a tribute to their endurance and their wherewithal to sustain one another's spirits that is so evident in their words for which we can be thankful there remains a record.

Conclusion

It has been my great good fortune to be able to immerse myself in the work required to produce two books both of which have been labors of love. The experience of researching and writing this book has not only enabled me to learn about an almost entirely forgotten chapter in U.S. and World War II history, a chapter that it is my hope will resonate with many readers, but it has made it possible to develop a most delightful friendship with the man whose determination to get the story of V-Mail out into the world has been so inspirational for me. Bill Streeter was a man of substance who for his entire adulthood devoted himself to sharing stories – of his beloved Cummington, Massachusetts in particular, but of our culture as evidenced in his brilliant, heartfelt preface.

As I have come to understand and appreciate the central role of letters in general and V-Mail in particular through this project, it has become clear to me that its component parts – microfilm and wartime communication – each deserve recognition in this conclusion. I will start with paying homage to microfilm by offering the reflections of another friend, author Howard Friel. It was on one of our morning walks that I asked Howard, who I knew to have done tremendous amounts of research for the acclaimed books he has written covering a broad range of subjects, from climate change to Israel, from Alzheimer's Disease to the reporting of *The New York Times*, about his microfilm experience.

I asked Howard what role microfilm played in any of his books. He responded by saying that he used microfilm "almost exclusively" for his first book on *The New York Times* called *The Record of the Paper: How The New York Times Misreports US Foreign Policy* which he wrote with Richard Falk (Verso, 2004). He told me that he enjoyed using microfilm because "you'd see things roll by that were of interest, whether they were relevant to the book or not," and that it "often took you back to a given era in interesting ways." He compared looking at the "antiseptic" headlines about the Vietnam War from *The New York Times* search engine of today to looking at the same headlines on microfilm. He described the latter telling me it "was like watching the TV series *Mad Men* about the world of advertising in the 1960s, with the display advertisements, ad copy, and photographs of the Vietnam War period, the pictures in the news stories, the stories themselves, the visual and print depictions of gender roles, and so on.

Howard acknowledged the trade-offs that favor the *Times*' search engine as follows:

> If you wanted to look something up in the microfilm version of *The New York Times*, you'd go to the library to locate *The New York Times* Index, which is a big, red reference book. And almost always you would need several volumes of the Index. You would search the several index volumes by date and subject, then write the dates and title of the articles in your notebook. Then you'd go to the microfilm room, look up the relevant role of microfilm among hundreds of others in big file-cabinet drawers that were chronologically arranged. Then you'd take the rolls to the microfilm machine, where you would manually scroll through several editions of the newspaper, scan the pages for what you were looking for, then read what you had found. You'd often print copies of articles from the microfilm machines to take with you. And these were temperamental machines that would occasionally fail to print, in which case you'd have to find someone in the library to fix it. And sometimes there wasn't anybody at the library to do that. Today, *The New York Times* search engine makes all of that unnecessary and so, at least if you are writing books about *The New York Times*, the process today takes much less time, all of which you can do from your home or office, but there is a price to pay for those efficiencies. Like walking to the library. Chatting with the familiar faces there. Sitting back

to look at the pages of the *Times* go by, and at the time that has gone by. So all of that is lost. And despite the frustrations, a lot of that was what made the process fun, and also more enlightening in some respects."

Howard concluded by saying that he didn't see a future for microfilm "because we've got Google and various specialized search engines that obviously work far more efficiently than putting on your coat and hat and lugging your briefcase over to the library to sit at the microfilm machine." At the same time, microfilm "is another bygone era that makes you wonder if the technological tradeoffs are worth it."

It is always worth noticing what we give up when our technology makes a quantum leap as it has from the world of microfilm to the digital age. First, a last word about V-Mail's effectiveness – a bit of a contradictory last word to be sure. The article on the V-Mail page of the http://alphabetilately.org/V.html website has this to say about V-Mail – its enormous space and weight saving advantages and its limited use to and from Navy personnel:

> The National Postal Museum website tells us that the space savings ratio of V-Mail was 37 to 1, while that for the weight was 57 to 1 – significant by either standard. On the other hand, only about 14 percent of mail to Navy personnel in 1944 was V-Mail, so while the system yielded significant savings, they could have been much larger if people had used it more.

And this:

> Between June 15, 1942 and April 1, 1945, 556,513,795 pieces of V-Mail were sent from the U.S. to military post offices and over 510 million pieces were received from military personnel abroad. In spite of the patriotic draw of V-Mail, most people still sent regular first class mail. In 1944, for instance, Navy personnel received 38 million pieces of V-Mail, but over 272 million pieces of regular first class mail.

What held people back from availing themselves of the "new technology," which chapter one clearly reveals was far from new. Starting in 1839 and continuing with its illustrious history onto the backs of pigeons during the 1870 Siege of Paris, microfilm has been a mainstay of our photographic and communication experiences. But old habits die hard. Airmail itself was not an old habit, since the first scheduled U.S. Air Mail service began in 1918, but many were accustomed to the idea of sending mail via airplanes. The subtle and, at times, not so subtle, pressure to avail one's self of V-Mail forms and comply with the accompanying rules were stumbling blocks that all of the encouragement – including detailed reasons why V-Mail was a patriotic response to having a loved one engaged in war – could not overcome. Those who used the system overlooked its hurdles, most particularly size and space limitations with the forms, by either simply accepting the limits, writing extra pages on additional forms (paying the additional postage if mailed from the U.S.) or writing and complaining. Nonetheless, it appears unquestionable that V-Mail (as well as Airgraph) had a role to play and it played it exceedingly well thanks to all of its component parts – from its originators to Kodak's contribution of equipment and know-how, to the companies that advertised its usage, to the countless men, women and children at home and abroad who wrote letters using V-Mail forms.

The final word will be about what has happened in the intervening years. Microfilm was supposed to have a most prolific future when the war ended its V-Mail function. For a time it most certainly dominated the research field in terms of making newspapers, magazines and periodicals readily available in library settings. It continued to serve its before-the-war function in the business world as well and Kodak was completely committed to its growth post-war. Here is a forecast of the future of microfilm from *Microfilm Systems – The First 40 Kodak Years* written in 1968:

> Progress has gone on apace and as microfilm attains its fortieth anniversary, there is virtually no line of endeavor or

undertaking which is at all concerned with records that is not in some way microfilm-oriented.[223]

Plans were clearly underway for the worlds of microfilm and computers to meet as these further announcements and prognostications indicate.

Anticipating the surge of the second forty years of microfilming (dating to its first use in banks in 1928), Recordak Corporation was merged with Eastman Kodak Company in January of 1965 and Recordak headquarters were moved from New York to Rochester. The merged Kodak-Recordak organizations anticipate a future in which the extraordinary capabilities of the computer will be paired with the equally extraordinary capabilities of microfilming in the specialized areas of documentation, push-button information retrieval, print-out and refilling – all at computer speeds. It looks to the elimination of diverting costly computer time to slow-speed mechanical print-out functions in favor of converting computer language on magnetic tape to plain language on microfilm at a speed of tens of thousands of characters per second. It contemplates a fast acceleration of microfilm applications and acceptance in world-markets. And it recognizes that, on the forty-year record of its contributions to the efficiency of communications, the advancement of systems technology, records management, education, and the preservation of man's records against disaster, microfilm is indeed here to stay and to flourish.[224]

'Twas not to be, despite these claims and hopes for microfilm contributing to and heading up technological advances.

But what about V-Mail? What about the impact of the written word as a means of maintaining meaningful, morale boosting connections? This is how the website author quoted above introduced his subject on the V-Mail page after illustrating what V-Mail made possible:

CONCLUSION

All of the above may seem pointless today, with cell phones and e-mail and virtually instantaneous communication to almost any point on the globe, but in the 1940s even air travel was still uncommon. No computers. No television. No space shuttle. No satellites. No velcro. How did people survive?

At least one answer during the war was using "photographic letters on wings." But how is someone who has grown up on "instantaneous communication" going to be able to begin to comprehend what it felt like to wait for mail call on the European or Pacific Fronts? How can a millennial imagine the excitement of seeing a V-Mail envelope drop into a mailbox of a wife, father, mother, sibling or child of a soldier? Our media is actually conspiring to make such visions into the past even more challenging. Take the film "Three Kings" as an illustrative example. The premise is that, "In the aftermath of the Persian Gulf War, four soldiers set out to steal gold that was stolen from Kuwait, but they discover people who desperately need their help." http://www.imdb.com/title/tt0120188/

The film was released in 1999, long before this book project was on my radar. And yet... the moment I saw the scene I am about to describe, I knew that warfare communication had arrived at another level entirely from anything that had come before.

The scene I refer to features Mark Wahlberg's character, Troy, having been captured, taken to a bunker and thrown into a room full of Kuwaiti cell phones. He is a prisoner of war, yet with any one of those phones he can live that dream depicted in the commercial to "reach out and touch" a loved one. He calls his wife back home who, needless to say, is both thrilled to hear his voice and completely panicked once she is told of his plight. If the situation could get any more bizarre, right after he tells her to report his location to his local Army Reserve unit, he is dragged away from the phone to be interrogated by the Iraqi officer in command of the bunker. Of course, his wife has no idea what his fate is to be and is left holding the disconnected phone in both disbelief and utter dread.

When I saw this scene I was overwhelmed with emotion. I felt the anguish of Troy's desperate situation and his decision to call home as his only means of possible succor. Competing with those emotions was my sense of his wife's helplessness having her husband reach out and being unable to help rescue him or even calm his distressed state.

Now when I think about such a moment, which can happen in similar circumstances during what is being viewed by many as America's perpetual wars, and I think about what was involved in writing, sending and receiving V-Mail, my mind reels at the contrast. From instantaneity and the accompanying chaos of war to having time to reflect, to ponder what one wants or needs to express, to know that your loved one will not receive your words for up to two weeks and could be some place completely different from where they were when you wrote the letter – could it be a more different reality?

Yet our species, which continues to resort to the ultimate experience of violence that is a war, has somehow succeeded, through force of will, technological advances and expediency to bridge the gap: to have soldiers and those who love them be able to write V-Mail letters that take weeks to arrive on the other side of the ocean or planet and that convey love, hope and faith that he or she is still alive and will return safely; to the immediacy of a cell phone call or a text that assures the loved one is still alive though the present moment could be life-threatening as is so often true in a war.

It is vital that we consider both what we want to communicate and how we want it to occur. V-Mail was an invaluable response to both of these considerations and I am very pleased to have had the opportunity to explore its history, its operation, its effectiveness and most movingly, the voices of those who chose to avail themselves – for the sake of their loved one and to contribute to the efforts to win the war – of its offerings. It is my hope that those who have learned the story of V-Mail will share what has been meaningful so this chapter of our history will receive its due.

Acknowledgments

There would be no book if not for several people who deserve to be acknowledged for their contributions to the book's creation. First and foremost, the inimitable Bill Streeter, Renaissance man par excellence whose numerous careers were always characterized by his dignity, kindness, grace, talent and extraordinary dedication to quality. Whether he was superbly binding books or writing an incomparable history of his beloved hometown, Cummington, Massachusetts, Bill pursued excellence, knowledge and a way to impart what he was learning and doing to a wide range of individuals who saw in him someone to emulate.

I had only known Bill since the summer of 2016. During that time he struggled with health issues and confronted his challenges with characteristic optimism, humor and resilience. In addition to sharing his voluminous research materials for the book he intended to write all about V-Mail, Bill also generously offered his reflections on the subject, his personal experience V-Mailing his beloved double cousin, Henry Streeter, and access for me to the Cummington Historical Society where he is deeply valued for his never-ending commitment to his hometown. The scrapbook Henry's stepmother put together that resides at the Historical Society was made available to me and I got a personal tour of the Cummington Tavern Museum. In addition I had the good fortune of exchanging books with Bill during our first encounter – mine being: *Called to Serve: Stories of Men and Women Confronted by the Vietnam War Draft* and his being, *My Friend Bill: The Live of a Restless Yankee*, a short biography that attempts to do justice to Bill's extraordinary life written by a fellow resident of his retirement home, Paul Schratter.

Of course, Bill contributed directly to the writing of the book by agreeing to provide the preface, which gets the book off to a flying start.

From there Bill also offered constructive criticism and tremendous encouragement in the form of his excitement that the book he had envisioned fifteen years ago was coming to fruition. I was greatly saddened by his recent death as our friendship was so new and so gratifying, but I was comforted with the knowledge that I had succeeded in getting a draft of the completed book to him so he could know that his vision had come to fruition for an homage to his cousin, Henry, and for a history of V-Mail, which had sustained their relationship despite the war that separated them.

Two Steve's get credit next. Steve O'Halloran, a close friend of Bill's, took it upon himself to find a way to get the book into print given what he knew about Bill's great enthusiasm for the project that he was no longer able to complete himself. Steve O'Halloren brought the idea for the project to the second Steve, Steve Strimer, who has the wherewithal to make the project operational. He is the publisher of Levellers Press, which enabled him to seek someone to write the book that he would commit to publishing upon its completion.

Steve approached me upon learning of my retirement from forty years of teaching third to sixth grade at the Smith College Campus School last summer. The two Steve's prepared a compelling sales pitch, which didn't even include how I could get to know and work with the simply wonderful Bill Streeter. That would come later, they told me. First I had to decide whether to accept their offer to take on the project.

Given the outstanding research material I could access, the possibility of working with someone as magical as Bill Streeter and the absolutely fascinating subject matter, it soon became the proverbial "no-brainer," which brings me to the next person deserving recognition – my wife, Susan. Her support, patience with the time commitment such an endeavor requires and encouragement when I needed "a shot in the arm" were invaluable throughout the process of creating this work.

I also want to give credit to the three friends who agreed to read the "Voices of V-Mail" chapter. I asked Steve Trudel, John Berkowitz and Kenneth Hahn to take the pulse of the chapter from their vantage points to determine which V-Mail letter excerpts were moving, powerful and evocative. I felt that I had gotten so engrossed in the material and had read so many letters as I decided which ones to excerpt, that having more objective, less involved readers would be beneficial. I am deeply grateful for their willingness to take on this role and for the thoughtful and meaningful feedback they offered.

Finally, I want to acknowledge all of the men, women and children who took it upon themselves to write V-Mail letters (Airgraph ones as well, though I did not have access to Airgraphs as I did to the volume of V-Mail letters Bill had obtained) from 1942–1945. I strongly believe their letter-writing endeavors aided the war effort in a myriad of ways, not least of which was enabling loved ones to be in even better touch than was otherwise available, which boosted morale on both the homefront and the war fronts, while simultaneously making more space available for critical supplies and manpower to be flown and shipped to where the war was occurring. Certainly acknowledging Henry's ultimate sacrifice and that of many other soldiers who received and sent V-Mail is part of this recognition. I am so very grateful for their voices.

ENDNOTES

Chapter 1

1 De Sola, MICROFILMING (Essential Books, 1944)

2 (http://www.srlf.ucla.edu/exhibit/text/BriefHistory.htm)

3 "Frederick Scott Archer," *British Journal of Photography*. 22 (773): 26 February 1875, pp. 102-104.

4 Peres, Michael R. *Focal Encyclopedia of Photography: Digital Imaging, Theory and Applications*. Focal Press. 2007, p. 124.

5 *Appletons' annual cyclopaedia and register of important events of the year: 1862*. New York: D. Appleton & Company. 1863. p. 186

6 *Exhibition of the Works of Industry of All Nations 1851. Reports by the Juries on the Subject in the Thirty Classes into which the Exhibition was Divided.* (London: John Weale, 1852).

7 Cobb, Aaron (2012). "Is John F. W. Herschel an Inductivist about Hypothetical Inquiry?" *Perspectives on Science*. 20: pp. 409–39.

8 Kent, A., Lancour, H. and Dally, J., *Encyclopedia of Library and Information Science: Volume 25*, Dekker, New York, 1972, p.231.

9 Sutton, Thomas, Worden, John and Low, S., *A Dictionary of Photography*. Thomas Sutton, John Worden. S. Low, Son, and Company, 1858

10 *The Photographic News*. London: Cassell, Petter and Galpin *(1858-1908)*

11 http://www.microscopy-uk.org.uk/mag/indexmag.html?http://www.microscopy-uk.org.uk/mag/artdec08/bs-slides3.html

12 https://en.wikipedia.org/wiki/Stanhope_lens

13 *The Photographic Journal* by the Royal Photographic Society of Great Britain, Jan. 15 1864.

14 *Newsletter of the Illinois State Archives & The Illinois State Historical Records Advisory Board* http://www.cyberdriveillinois.com/publications/pdf_publications/fortherecordwinter07.pdf

15 *Chronology of Microfilm Developments 1800 – 1900* from UCLA, http://www.srlf.ucla.edu/exhibit/text/Chronology.htm

16 *The Strad Magazine* October 2005 pp. 51-54

17 https://en.wikipedia.org/wiki/René_Dagron

18 Luther, Frederick. *Microfilm: A History 1839 – 1900*. Barre, Mass. Barre Publishing Company. 1959.

19 https://en.wikipedia.org/wiki/Microfilm_reader

20 Levi, Wendell (1977). *The Pigeon.* Sumter, South Carolina: Levi Publishing Co, Inc.

21 Blume, Mary. "The hallowed history of the carrier pigeon." *The New York Times.* January 30, 2004. Accessed February 6, 2012. Online.

22 Ferguson, Niall. The House of Rothschild vol. 1. London: Penguin, 2000. Print.

23 Hayhurst, J.D. O.B.E.,*The Pigeon Post into Paris 1870-1871*

prepared in digital format by Mark Hayhurst Copyright 1970. (http://www.cix.co.uk/~mhayhurst/jdhayhurst/pigeon/pigeon.html)

24 Horne, Alistair. *The French Army and Politics 1870-1970*, Peter Bedrick Books, 1984. pg. 7.

25 Hayhurst, J.D. O.B.E.,*The Pigeon Post into Paris 1870-1871*, prepared in digital format by Mark Hayhurst Copyright 1970 John Hayhurst (http://www.cix.co.uk/~mhayhurst/jdhayhurst/pigeon/pigeon.html)

26 IBID.

27 IBID.

28 IBID.

29 IBID.

30 IBID.

31 IBID.

32 https://en.wikipedia.org/wiki/Pigeon_post

33 https://postalmuseum.si.edu/victorymail/introducing/photo1_microfilm.html

34 http://www.intercom.org.br/papers/nacionais/2014/resumos/R9-2793-1.pdf

35 Luther, Frederick. *Microfilm: A History 1839 – 1900.* Barre, Mass. Barre Publishing Company. 1959, p, 7.

36 Lawrence, Ashley, *A Message brought to Paris by Pigeon Post in 1870-71* UK, www.microscopy-uk.org.uk/mag/artoct10/al-pigeonpost.html

37 IBID.

38 IBID.

39 Hayhurst, J.D. O.B.E.,*The Pigeon Post into Paris 1870-1871,* op. cit.

40 Lawrence, Ashley, A Message brought to Paris by Pigeon Post in 1870-71 UK, www.microscopy-uk.org.uk/mag/artoct10/al-pigeonpost.html

41 https://en.wikipedia.org/wiki/Arc_lamp

42 http://www.srlf.ucla.edu/exhibit/text/hist_page4.htm

43 Lawrence, Ashley, A Message brought to Paris by Pigeon Post in 1870-71 UK, www.microscopy-uk.org.uk/mag/artoct10/al-pigeonpost.html

44 IBID.

45 IBID.

46 http://www.kodak.com/corp/aboutus/heritage/georgeeastman/default.htm

47 IBID.

48 Meckler, Alan M. (1982). *Micropublishing: A History of Scholarly Micropublishing in America, 1938–1980.* Westport, CT: Greenwood Press.

49 *Goldschmidt, Robert. "Sur une forme nouvelle du livre :le livre microphotographique. Bruxelles". Institut international de bibliographie.* Retrieved January 15, 2014

50 https://en.wikipedia.org/wiki/Robert_Goldschmidt

51 De, Sola R. *Microfilming.* New York: Essential Books, 1944. Print

52 Hilton, Frank L., Jr., *Microfilm Systems – the First Forty Years*, Kodak Library, Rochester, NY, 1968, p. 117.

53 IBID.

54 IBID.

55 IBID.

56 IBID, p. 118.

57 IBID, p.118.

58 IBID, p. 118.

59 http://americanarchivist.org/doi/pdf/10.17723/aarc.58.1.h715312706n38822

60 "*Brief History of Microfilm*," Heritage Microfilm, 2007, http://www.microfilmworld.com/briefhistoryofmicrofilm.aspx

61 https://www.reference.com/government-politics/difference-between-microfiche-microfilm-26d39a34663cada9

62 https://en.wikipedia.org/wiki/Microform

63 Hilton, Frank L., Jr., Microfilm Systems – the First 40 Years, Kodak Library, Rochester, NY, p. 119.

64 IBID, p. 119.

65 IBID, p. 120.

66 IBID, p. 123.

67 http://www.atlasobscura.com/articles/the-strange-history-of-microfilm-which-will-be-with-us-for-centuries

68 IBID.

69 IBID.

70 IBID.

71 https://www.tsl.texas.gov/slrm/blog/2010/10/why-do-we-still-need-microfilm/

72 from https://en.wikipedia.org/wiki/Microform

Chapter 2

73 Beam, Christopher. (2007-01-05) *How International Mail Works.* http://www.slate.com/articles/news_and_politics/explainer/2007/01/i_mailed_a_letter_to_paris_.html

74 https://en.wikipedia.org/wiki/Universal_Postal_Union

75 http://www.upu.int/en/the-upu/history/about-history.html

76 Beam, Christopher. (2007-01-05) How international mail works. Slate.com. Retrieved on 2014-04-28.

77 http://www.bbc.com/news/magazine-25934407

78 IBID.

79 IBID.

80 Campbell-Smith, Duncan, *Masters of the Post: The Authorized History of the Royal Mail,* Penguin Global; 1st Printing edition, 2012.

81 http://a.files.bbci.co.uk/bam/live/content/z2dtgk7/transcript

82 IBID.

83 http://www.bbc.com/news/magazine-25934407

84 *Between Front Line and Home,* http://encyclopedia.1914-1918online.net/article/war_letters_communication_

85 IBID.

86 https://en.wikipedia.org/wiki/Imperial_War_Museum

87 http://encyclopedia.1914-1918-online.net/article/war_letters_communication_between_front_and_home_front

88 http://www.bbc.com/news/magazine-25934407

89 http://www.club-together.org/lifestyle/articles/from-the-frontline-...with-love.aspx#.WF12krGZORs

90 http://www.bbc.co.uk/guides/zqtmyrd

91 https://en.wikipedia.org/wiki/Harold_Balfour,_1st_Baron_Balfour_of_Inchrye

92 http://alphabetilately.org/airgraph.html

93 http://stampomania.blogspot.com/2009/08/history-of-airgraph-from-india.html

94 Stephen, J., *Airgraph and V..._Mail Catalogue 1948*, George J. Schultze, Philadelphia, PA

95 IBID.

96 IBID.

97 http://alphabetilately.org/airgraph.html

98 Stepehen, J., op. cit.

99 IBID.

100 IBID.

101 IBID.

102 IBID.

103 IBID.

104 IBID.

105 IBID

106 IBID.

107 IBID.

108 IBID.

Chapter 3

109 Snyder, Lt. Colonel E.D., Radio News, February 1944.

110 IBID.

111 IBID.

112 IBID.

113 IBID.

114 IBID.

115 McCarthy, George L., *Internal Memo*, Kodak Corp., 1942.

116 IBID.

117 "V-Mail on Motion Picture Film (microfilm) is Army's Plan to Expedite Soldier Mail." *War Department Immediate Press Release*, March 5, 1942.

118 Rose, Brigadier General William. *War Department Immediate Press Release,* May 21, 1942.

119 Snyder, Lt. Colonel E.D. , *Radio News*, February 1944.

120 Hudson, James W., "Funny Mail," Newsletter of *The American Philatelist*, July 1994, p. 608.

121 IBID.

122 "Roosevelt Gets Two Messages Opening Overseas Service V-Mail," *The New York Times*, June 11, 1942.

123 IBID.

124 Hudson, James W., op. cit.

125 Kane, Joseph Nathan, *Famous First Facts: A Record of First Happenings, Discoveries and Inventions in the U.S.*, 7th Edition, H. W. Wilson Company, NY, 2015, page 471.

126 http://alphabetilately.org/V.html

127 Air Mail Entire Truth, #4714, N.Y. Lava Inc., 1963-1980.

128 Sherman, Lawrence, M.D. *The United States Post Office in World War II*, 2002.

129 https://postalmuseum.si.edu/collections/object-spotlight/V-Mail-letter-sheets.html

130 Bill Streeter Memorial V-Mail Research Material Archives

131 http://postalmuseum.si.edu/VictoryMail/using/index.html

132 Courtesy Rare Book & Texana Collections, University of North Texas Libraries. https://texashistory.unt.edu/explore/collections/UNTRB/

133 http://alphabetilately.org/V.html

134 "*The Type of Information Servicemen Prefer in Their Mail,* Office of War Department, May 27, 1943.

135 IBID.

136 IBID.

137 IBID.

138 IBID.

139 http://postalmuseum.si.edu/VictoryMail/letter/index.html

140 Ames, Rosemary, *Sabotage Women of America*; File E-NC-148-57/181; OWI Intelligence Digests, Office of War Information, Record Group 208; National Archives at College Park, Maryland; pp. 4-5.

141 McDonagh, Edward, and McDonagh, Louise, *War Anxieties of Soldiers and Their Wives*, Vol. 24, No. 2 (Dec., 1945), pp. 195-200

Oxford University Press: Oxford, England.

142 IBID, p. 197.

143 IBID, p. 200.

144 *V-Mail – A Good Medicine.*

145 *Shortage of Cargo Space Curtails Air Mail Service to Overseas Points*," War Department Immediate Press Release, March 5, 1945.

146 Smithsonian Institution, National Postal Museum.

147 http://www.pbs.org/wgbh/americanexperience/features/general-article/warletters-censorship/

148 National Archives (111-SC-142-170365-B).

149 National Archives (111-SC-818-384767).

150 National Archives (111-SC-142-170366-B).

151 http://alphabetilately.org/V.html

152 IBID.

153 *The Story of V-Mail,* Recordak Record, August 1959.

154 *Kodakery*, Newspaper for the Men and Women of Eastman Kodak Company, Rochester, NY, February 29, 1944.

155 *Army Postal Service During World War II*, p. 229.

156 IBID, p. 229-230.

157 "V-Mail Serves Useful Wartime Purpose," *Linn's Stamp News*, Monday, November 10, 1975.

158 IBID.

159 *Business Week Magazine*, February 1945, p. 40.

160 "100,000,000 Times Soldiers of the European Theater Have Used V-Mail to Send a Message Home," *The New York Times*, Saturday, May 27, 1944.

161 https://postalmuseum.si.edu/victorymail/introducing/how.html

Chapter 4

162 Clark, Dr. Walter, Kodak Research Laboratories in April, 1943.

163 Snyder, Lt. Colonel E.D. *V-Mail: The adoption of V-Mail service has reduced shipping tonnage to overseas ports.*

164 IBID.

165 IBID.

166 IBID.

167 "Will Bar Lipstick V-Mail – Postal Officers Set Deadline at St. Valentine's Day," *The New York Times* on February 4, 1944.

168 Snyder, Lt. Colonel E.D. *V-Mail: The adoption of V-Mail service has reduced shipping tonnage to overseas ports.*

169 IBID.

170 IBID.

171 IBID.

172 IBID.

173 IBID.

174 IBID.

175 Campbell, Bruce J., "Photos Can Be Sent By V-Mail", *San Francisco Chronicle*, May 26, 1943.

176 IBID.

177 IBID.

178 IBID.

179 IBID.

180 Burlingame, Lieutenant Wes, Navy Combat Photo Unit.

181 *The New York Times*, June 15, 1944.

182 Snyder, Lt. Colonel E.D. *V-Mail: The adoption of V-Mail service has reduced shipping tonnage to overseas ports.*

183 *The History and Progress of Modern Microfilming by Recordak Corporation*, Subsidiary of Eastman Kodak Company, 350 Madison Ave., NY.

184 IBID.

Chapter 5

185 "V-Mail – A Good Medicine", Bill Streeter Memorial V-Mail Research Material Archives

186 *Forthnightly Budget: For Wartime Editors of Women's Pages*, Office of War Information, February 5, 1944.

187 *The New York Times*, November 7, 1943.

188 Howard, Ronald, *In Search of My Father: A Portrait of Leslie Howard*, St. Martin's Press, New York 1981.

189 "The Airmail Entire Truth" newsletter, No. 11, May 1964.

190 *The New York Times*, September 1944.

191 "African Belles Aid V-Mailers," *Kodakery*, September 1943.

192 IBID.

193 *Kodakery*, September 16, 1943.

194 "V-Mail – A Good Medicine", Bill Streeter Memorial V-Mail Research Material Archives

195 IBID.

196 IBID.

197 IBID.

198 *Kodakery*, November 9, 1943.

199 IBID.

200 "Mail Unit Lands Quickly," *The New York Times*, February 15, 1944.

201 *Kodakery*, June 20, 1944.

202 IBID.

203 IBID.

204 IBID.

205 "V-Mail – A Good Medicine", Bill Streeter Memorial V-Mail Research Material Archives

206 *V-Pac News*, *Kodakery*, DATE NEEDED?

207 *Linn's Stamp News*, Monday, November 10, 1975.

208 *The Local Council Exchange*, February 1945 edition of Boy Scout publication.

209 *Kodakery*, December 12, 1944.

210 *Kodakery*, October 10, 1944.

211 *Kodakery*, February 27, 1945.

Chapter 6

212 http://scriptorium.lib.duke.edu/adaccess/vmail.html.

213 http://alphabetilately.org/V.html

214 IBID.

215 IBID.

216 IBID.

217 IBID.

218 Reep. Edward, *Correspondence with officer serving during World* War II and interview subject for the research for this project, 2002.

219 IBID.

220 IBID.

221 IBID.

222 *The New York Times*, June 18, 1944

CONCLUSION

223 Hiliton, Frank L., *Microfilm Systems – The First 40 Years*, Kodak Library, Rochester, NY, 1968, p. 125.

224 IBID.

BIBLIOGRAPHY

PRINT MATERIALS

"African Belles Aid V-Mailers," *Kodakery*, September 1943.

The Airmail Entire Truth Newsletter, #4714, N.Y. Lava Inc., 1963-1980.

The Airmail Entire Truth Newsletter, No. 11, May 1964.

Ames, Rosemary, *Sabotage Women of America*; File E-NC-148-57/181; OWI Intelligence Digests, Office of War Information, Record Group 208; National Archives at College Park, Maryland.

Army Postal Service During World War II.

Appletons' annual cyclopaedia and register of important events of the year: 1862. New York: D. Appleton & Company. 1863.

Blume, Mary. "The hallowed history of the carrier pigeon." *The New York Times.* January 30, 2004. Accessed February 6, 2012.

Burlingame, Lieutenant Wes, Navy Combat Photo Unit.

Business Week Magazine, February 1945.

Campbell, Bruce J., "Photos Can Be Sent By V-Mail", San Francisco Chronicle, May 26, 1943.

Campbell-Smith, Duncan, *Masters of the Post: The Authorized History of the Royal Mail,* Penguin Global; 1st Printing edition, 2012.

Clark, Dr. Walter, Kodak Research Laboratories in April, 1943

Cobb, Aaron (2012), "*Is John F. W. Herschel an Inductivist about Hypothetical Inquiry?*" Perspectives on Science. 20.

De Sola, *Microfilming,* (Essential Books, 1944).

Exhibition of the Works of Industry of All Nations 1851. Reports by the Juries on the Subject in the Thirty Classes into which the Exhibition was Divided. (London: John Weale, 1852).

Ferguson, Niall, *The House of Rothschild* vol. 1. London: Penguin, 2000.

Forthnightly Budget: For Wartime Editors of Women's Pages, Office of War Information, February 5, 1944.

"Frederick Scott Archer," *British Journal of Photography*. 22 (773): 1875.

Goldschmidt, Robert. "Sur une forme nouvelle du livre :le livre micropho-tographique. Bruxelles". *Institut international de bibliographie*. Retrieved January 15, 2014.

Hilton, Frank L., Jr., *Microfilm Systems – the First Forty Years*. Kodak Library, Rochester, NY.

Howard, Ronald, *In Search of My Father: A Portrait of Leslie Howard*, St. Martin's Press, New York 1981.

Hudson, James W., "Funny Mail," Newsletter of *The American Philatelist*, July 1994.

Kane, Joseph Nathan, *Famous First Facts: A Record of First Happenings Discoveries and Inventions in the U.S.*, 7th Edition, H.W. Wilson Company, NY, 2015.

Kent, A., Lancour, H. and Dally, J., *Encyclopedia of Library and Information Science: Volume 25*, Dekker, New York, 1972, p.231.

Kodakery, Newspaper for the Men and Women of Eastman Kodak Company, Rochester, NY, February 29, 1944.

Kodakery, September 16, 1943.

Kodakery, November 9, 1943.

Kodakery, June 20, 1944.

Kodakery, December 12, 1944.

Kodakery, October 10, 1944.

Kodakery, February 27, 1945.

V-Pac News, Kodakery, date unknown

Levi, Wendell,*The Pigeon*. Sumter, South Carolina: Levi Publishing Co, Inc., 1977

Linn's Stamp News, Monday, November 10, 1975.

Luther, Frederick, *Microfilm: A History 1839 – 1900*. Barre, Mass.

"Mail Unit Lands Quickly," *The New York Times*, February 15, 1944. Publishing Company. 1959.

McCarthy, George L., *Internal Memo*, Kodak Corp., 1942

McDonagh, Edward, and McDonagh, Louise, *War Anxieties of Soldiers and Their Wives*, Vol. 24, No. 2 (Dec., 1945), pp. 195-200, Oxford

University Press: Oxford, England.

Meckler, Alan M. (1982), *Micropublishing: a history of scholarly micropublishing in America, 1938–1980.* Westport, CT: Greenwood Press.

National Archives, (111-SC-142-170365-B).

National Archives, (111-SC-818-384767).

National Archives, (111-SC-142-170366-B).

The New York Times, June 15, 1944.

The New York Times, September 1944.

The New York Times, June 18, 1944

"100,000,000 Times Soldiers of the European Theater Have Used V-Mail to Send a Message Home," *The New York Times*, Saturday, May 27, 1944.

Peres, Michael R. *Focal Encyclopedia of Photography: Digital Imaging, Theory and Applications.* Focal Press. 2007, p. 124.

Reep, Edward, *Correspondence with officer serving during World* War II and interview subject for the research for this project, 2002.

"Roosevelt Gets Two Messages Opening Overseas Service V-Mail," *The New York Times*, June 11, 1942

Rose, Brigadier General William, War Department Immediate Press Release, May 21, 1942.

Sherman, Lawrence, M.D., *The United States Post Office in World War II*, 2002.

"Shortage of Cargo Space Curtails Air Mail Service to Overseas Points," War Department Immediate Press Release, March 5, 1945.

Snyder, Lt. Colonel E.D., Radio News, February 1944.

Stephen, J., *Airgraph and V..._Mail Catalogue 1948*, George J. Schultze, Philadelphia, PA.

Sutton, Thomas, Worden, John and Low, S., *A Dictionary of Photography.* Thomas Sutton, John Worden. S. Low, Son, and Company, 1858.

The Local Council Exchange, February 1945 edition of Boy Scout publication.

The History and Progress of Modern Microfilming by Recordak Corporation, Subsidiary of Eastman Kodak Company, 350 Madison Ave., NY.

The Photographic Journal, Royal Photographic Society of Great Britain, Jan. 15 1864.

The Photographic News. London: Cassell, Petter and Galpin, 1858-1908.

The Story of V-Mail, Recordak Record, August 1959. *V-Mail – A Good Medicine*

"The Type of Information Servicemen Prefer in Their Mail," Office of War Department, May 27, 1943.

V-Mail on Motion Picture Film (microfilm) is Army's Plan to Expedite Soldier Mail. War Department Immediate Press Release, March 5, 1942.

"V-Mail Serves Useful Wartime Purpose," *Linn's Stamp News*, Monday, November 10, 1975.

"Will Bar Lipstick V-Mail – Postal Officers Set Deadline at St. Valentine's Day," *The New York Times*, February 4, 1944.

World War II Magazine. Published in May 1951

DIGITAL SOURCES

http://www.microscopy-uk.org.uk/mag/indexmag.html?http://www.microscopy-uk.org.uk/mag/artdec08/bs-slides3.html

https://en.wikipedia.org/wiki/Stanhope_lens

Newsletter of the Illinois State Archives & The Illinois State Historical Records Advisory Board, http://www.cyberdriveillinois.com/publications/pdf_publications/fortherecordwinter07.pdf

Chronology of Microfilm developments 1800 – 1900 from UCLA, http://www.srlf.ucla.edu/exhibit/text/Chronology.htm

http://www.thestrad.com/downloads/PictureBows.pdf

https://en.wikipedia.org/wiki/René_Dagron

BIBLIOGRAPHY

Hayhurst, J.D. O.B.E., *The Pigeon Post into Paris 1870-1871* prepared in digital format by Mark Hayhurst Copyright 1970. (http://www.cix.co.uk/~mhayhurst/jdhayhurst/pigeon/pigeon.html)

https://postalmuseum.si.edu/victorymail/introducing/photo1_microfilm.html

https://en.wikipedia.org/wiki/Pigeon_post

http://www.intercom.org.br/papers/nacionais/2014/resumos/R9-2793-1.pdf

Lawrence, Ashley, *A Message brought to Paris by Pigeon Post in 1870-71* UK, www.microscopy-uk.org.uk/mag/artoct10/al-pigeonpost.html

http://www.kodak.com/corp/aboutus/heritage/georgeeastman/default.htm

https://en.wikipedia.org/wiki/Robert_Goldschmidt

http://americanarchivist.org/doi/pdf/10.17723/aarc.58.1.h715312706n38822.

"*Brief History of Microfilm,*" Heritage Microfilm, 2007, http://www.microfilmworld.com/briefhistoryofmicrofilm.aspx

https://www.reference.com/government-politics/difference-between-microfiche-microfilm-26d39a34663cada9

https://en.wikipedia.org/wiki/Microform

http://www.atlasobscura.com/articles/the-strange-history-of-microfilm-which-will-be-with-us-for-centuries

"*Brief History of Microfilm,*" Heritage Microfilm, 2007, http://www.microfilmworld.com/briefhistoryofmicrofilm.aspx

https://www.reference.com/government-politics/difference-between-microfiche-microfilm-26d39a34663cada9

https://www.tsl.texas.gov/slrm/blog/2010/10/why-do-we-still-need-microfilm/

Beam, Christopher. (2007-01-05) *How International Mail Works.* http://www.slate.com/articles/news_and_politics/explainer/2007/01/i_mailed_a_letter_to_paris_.html

https://en.wikipedia.org/wiki/Universal_Postal_Union

http://www.upu.int/en/the-upu/history/about-history.html

http://www.bbc.com/news/magazine-25934407

http://a.files.bbci.co.uk/bam/live/content/z2dtgk7/transcript

Between Front Line and Home, http://encyclopedia.1914-1918online.net/article/war_letters_communication_

https://en.wikipedia.org/wiki/Imperial_War_Museum

http://www.club-together.org/lifestyle/articles/from-the-frontline-…with-love.aspx#.WF12krGZORs

http://www.club-together.org/lifestyle/articles/from-the-frontline-…with-love.aspx#.WF12krGZORs

http://www.bbc.co.uk/guides/zqtmyrd

https://en.wikipedia.org/wiki/Harold_Balfour,_1st_Baron_Balfour_of_Inchrye

http://alphabetilately.org/airgraph.html

http://stampomania.blogspot.com/2009/08/history-of-airgraph-from-india.html

https://postalmuseum.si.edu/collections/object-spotlight/V-Mail-letter-sheets.html

http://postalmuseum.si.edu/VictoryMail/using/index.html

Courtesy Rare Book & Texana Collections, University of North Texas Libraries. https://texashistory.unt.edu/explore/collections/UNTRB/

http://www.pbs.org/wgbh/americanexperience/features/general-article/warletters-censorship/

https://postalmuseum.si.edu/victorymail/introducing/how.html